FINANCIAL RISK AND DERIVATIVES

A Special Issue of the Geneva Papers on Risk and Insurance Theory

edited by Henri Loubergé
and Marti G. Subrahmanyam

Reprinted from the
Geneva Papers on Risk and Insurance Theory
Volume 21, Number 1

Kluwer Academic Publishers
Boston/Dordrecht/London

The Geneva Papers on Risk and Insurance Theory

Volume 21, No. 1, June 1996

Special Issue on Insurance and Financial Risk Management
Guest Editors: Henri Loubergé and Marti G. Subrahmanyam

Distributors for North America:
Kluwer Academic Publishers
101 Philip Drive
Assinippi Park
Norwell, Massachusetts 02061 USA

Distributors for all other countries:
Kluwer Academic Publishers Group
Distribution Centre
Post Office Box 322
3300 AH Dordrecht, THE NETHERLANDS

Library of Congress Cataloging-in-Publication Data

A C.I.P. Catalogue record for this book is available from
the Library of Congress.

Printed on acid-free paper.

Printed in the United States of America

The Geneva Papers on Risk and Insurance Theory, 21: 5-6 (1996)

Introductory Note

HENRI LOUBERGÉ
University of Geneva, Department of Economics, 102 Blvd. Carl Vogt, CH-1211 Geneva 4, Switzerland

This *Special Issue on Insurance and Financial Risk Management* is composed of articles selected from the program of the Twenty-Second Seminar of the European Group of Risk and Insurance Economists, which was held at the University of Geneva in September 1995. It provides a nice illustration of the links that have developed over the past years between the theory of finance on one hand and insurance economics and actuarial science on the other hand. Advances in contingent claims analysis and developments in the academic and practical literature dealing with the management of financial risks reflect the close relationships between insurance and innovations in finance.

The first article in the issue, on models of the term structure of interest rates, is based on the 1995 Geneva Risk Lecture, delivered by Marti Subrahmanyam. It provides a comprehensive review of the issues involved in modeling the term structure of interest rates and valuing contingent claims on interest-rate products. The author focuses on the central issues, pointing out the questions that were answered satisfactorily so far and the gaps remaining in our understanding, without leading the reader into the technical details of the models. He provides an illuminating overview of five seminal models that laid the foundations for further work in this area. Each model represents a specific approach to term structure modeling, and Marti is particularly careful in stressing the strengths and weaknesses of each approach, as well as their similarities and differences. His article will certainly become a key reference for future work on the term structure of interest rates, both as an introductory reading and as an anchor to keep research focused on central issues.

The following article, by Günter Franke, brings some remarks on Marti's lecture by replacing work on the term structure into the broader perspective of capital market research, and deriving some implications for financial risk management. In this respect, the author stresses that the use of different models for different purposes may lead to illusions of arbitrage opportunities. Moreover, the need to monitor the risk position of financial intermediaries frequently (to allow for pricing errors) lead to short-termism among these institutions.

The two following articles deal with financial innovation in life insurance. First, Steinar Ekern and Svein-Arne Persson analyze exotic unit-linked life insurance contracts. These products with payoffs linked to the return on a financial asset are called "exotic" because various arrangements possible: guarantees, cap on benefits, links to several financial assets or portfolios of assets, and so on. The contracts are also characterized by embedded options, such as a conversion option held by the insured, a buy-back option held by the insurer, the option to choose the financial asset used for the link, and so forth. The patterns of payoffs are similar to those of exotic options available on derivative markets. The two authors are therefore able to draw on results in the financial literature to price these contracts. In particular, they illustrate how to compute appropriate single premiums as market values of the unit-linked contracts, combining financial and mortality risks. Then, Jörg-Aase Nielsen and Klaus Sandmann focus on the uniqueness of the fair premium for equity-linked life insurance contracts. They combine actuarial and financial theory to price numerically a specific life insurance product. This product is characterized by periodic premium payments and a bonus on top of the guaranteed amount, which is similar to an Asian option with exercise price equal to the guaranteed amount. Three risks are taken into account in the model: mortality risk (diversifiable in the insurer's portfolio), interest-rate risk, and the price risk of the asset to which the contract is linked. It turns out that the fair periodic premium is an increasing and convex function of the share invested in the financial asset.

The next article, written by Abraham Lioui, Pascal Nguyen Duc Trong, and Patrice Poncet, deals with optimal dynamic hedging in incomplete futures markets. It is not focused on an insurance problem, like the two preceding articles, since its objective is to provide a contribution to the general financial literature on hedging. The authors consider a logarithmic utility investor endowed with an exogenous portfolio of risky assets. The investor is not allowed to do any trading in the spot market, but she may hedge by trading in futures contracts expiring at her investment horizon T. Given that the market is incomplete, the dynamic hedging policy is solved by exploiting a methodology utilizing equivalence between particular static and dynamic programs. The same problem is also solved for a CARA investor under the assumption of lognormal price increments.

The last article in the issue, written by Suk Heun Yoon and Sumon Mazumdar, deals with the important financial policy issues of deposit insurance and banking regulation. The objective is to design a regulation that prevents banks from taking excessive risks in their supply of credit. The article assumes informational asymmetries between the regulator and the bank regarding loan quality. It is shown that, assuming the regulator can observe actual rates charged by the banks to their customers, imposing a capital requirement that links the unobserved loan quality to the loan rate is incentive compatible: competitive banks are indifferent to all loans satisfying a minimum acceptable quality and reject all riskier loans. Hence, the regulator's task is to design a capital requirement schedule that leads loan rates to be decreasing in loan quality. The conditions for reaching this goal are defined in the article. As the authors remark, their result provides theoretical support for suggested policy reforms in which banks are required to put up more equity capital on riskier assets.

In summary, the articles published in this issue deal with different subjects and use different methodological approaches. They represent an overview of the present state of the art in theoretical research dealing with financial issues of significance for insurance science. They will hopefully provide an impetus to further developments in applied insurance research.

Acknowledgments

The guest editors of this Special Issue would like to express their gratitude to the following referees for their help in the preparation of the issue: Knut Aase, Eric Briys, Stephen Diacon, Steinar Ekern, Benton Gup, Philip Hardwick, Yaffa Machnes, Sumon Mazumdar, Jörg-Aase Nielsen, Svein-Arne Persson, Auke Plantinga, and Patrick Roger.

The Geneva Papers on Risk and Insurance Theory, 21: 7–28 (1996)

The Term Structure of Interest Rates: Alternative Approaches and Their Implications for the Valuation of Contingent Claims

MARTI G. SUBRAHMANYAM
Leonard N. Stern School of Business, New York University, 44 West Fourth Street, New York, NY 10012-1126

Abstract

One of the most active areas of research in financial economics has been the modeling of the term structure of interest rates and its relationship to the pricing of contingent claims. There is a vast array of issues in the area, as well as a variety of perspectives, ranging from theoretical to practical. This article provides a general framework for the analysis of issues in the modeling of the term structure. Specifically, this article provides an overview of the conceptual issues and the empirical evidence in the area, based on an examination of five seminal models by Black, Scholes, and Merton; Vasicek; Cox, Ingersoll, and Ross; Ho and Lee; and Heath, Jarrow, and Morton. The article provides a synthesis of the area and suggests directions for future research.

Key words: term structure, interest rates, contingent claims valuation, Black-Scholes model, mean-reversion, no-arbitrage condition, preference-free pricing, general equilibrium, equivalent martingale measure

1. Introduction

One of the most active areas of research within the literature on financial economics in recent years has been the modeling of the term structure of interest rates and its relationship to the pricing of interest-rate contingent claims. The interest in this area stems both from the conceptual richness of the problems posed, as well as the practical issues that arise in the implementation of the models developed. It goes without saying that a considerable impetus for this research, particularly on questions of implementation, has come from the world of practice, which has experienced a virtual explosion in the variety and quantity of interest-rate derivative products.

A casual survey of the area suggests that researchers have addressed a variety of issues in the modeling of the term structure, often in the context of particular models. These issues span the spectrum from the equilibrium determination of interest rates to alternative specifications of interest-rate processes and from arbitrage-free pricing models for bonds and interest-rate contingent claims to the implementation of these models. Often, the papers in the area are concerned with both the modeling of such claims with a focus on pricing issues from a financial economist's perspective, as well as a solution to problems of implementation from a financial engineering point of view. These diverse perspectives sometimes make it difficult to place the work in the area in a broader framework.

Given the vast array of issues in the area, as well as the variety of perspectives, ranging from a theoretical financial economist's to that of a practitioner in the field, there is an amazing quantity of detail that confronts anyone who attempts to get an overall perspective of the field. Not surprisingly, perhaps, there have been a number of surveys of the literature in the area. One problem with these surveys is that while many of them catalog the various papers in the area, sometimes in a fair amount of detail, it is difficult to get a grip on what the central issues are, what questions have been answered satisfactorily, and where the gaps are in our understanding. In addition, the distinction is not often made between the general requirements for such models, such as an arbitrage-free term structure, that have to be satisfied by *any* model that is proposed, and issues of detail, such as the precise specification of the stochastic processes, which have to be resolved ultimately by empirical examination.

This article attempts to provide such an analysis but from an overall perspective rather than from focusing on questions of detail. Two issues of general interest in the area of term structure modeling and the valuation of interest-rate contingent claims will not be discussed here in much detail: problems of implementation of the models and the empirical evidence. Each of these issues requires a detailed examination in its own right, and it would be overly ambitious to survey them in the context of the present general overview of the literature. However, the main conclusions of the empirical evidence are presented here as well as the highlights of the implementation issues.

Section 2 discusses the main issues that have been posed in the broad area of term structure modeling and places them in context. Section 3 provides an overview of five seminal models in the area that provide the foundation for further work. A brief review of the empirical issues in the area and a summary of the main conclusions are given in Section 4. In the next section, Section 5, an attempt is made to synthesize the work by defining issues where there appears to be general agreement. Section 6 concludes by discussing the issues that remain unresolved and possible directions for further work in the area.

2. Issues in modeling the term structure of interest rates and the pricing of interest-rate contingent claims

At a general level, financial economists are concerned with the equilibrium foundations of the model they develop. Although most financial models are pure-exchange models and are not concerned with production economies, it is important to ensure that there is no inconsistency between the asset return processes that are assumed and the underlying production economy. In particular, the question is how the state variables in the underlying production economy, when placed in equilibrium setting, yield the commonly used specifications of the stochastic processes generating asset returns.

The issue of consistency between the underlying production economy and the stochastic process generating asset returns is virtually impossible to answer at a completely general level. Since the equilibrium prices of assets are determined by the interaction of the preferences of consumer-investors and the production processes that are available, it would be necessary to specify both assumptions in some detail in order to get an explicit analytical solution for the stochastic process for asset prices.[1]

Since most financial models deal only with pure exchange equilibria, they are partial equilibrium models viewed from the above perspective. However, even within such partial equilibrium models, the issue of consistency arises. In many cases, this amounts to a requirement that assets be priced consistently with respect to each other. This requirement can be stated as an arbitrage condition within the model—for instance, that pricing of different assets should imply the same market price of risk, i.e., the model has to be internally arbitrage-free. Again, it is difficult to derive such specific pricing results for arbitrary stochastic processes. Restrictions have to be placed on the form of the stochastic process in order to derive closed-form solutions for bond prices and the prices of contingent claims. Unfortunately, some processes that have the virtue of analytic tractability may have undesirable economic (or even econometric) properties. For instance, Gaussian interest-rate processes yield closed-form solutions for the prices of bonds and contingent claims but allow for negative nominal interest rates, which would be inconsistent with basic arbitrage requirements. From an econometric perspective, the choice of a process with no known unconditional distribution makes the use of estimation methods other than the generalized method of moments (GMM) hard to employ.

A basic requirement of contingent-claim models since the work of Black and Scholes [1973] is that the price of the underlying asset be taken as given and also that its current market price be consistent with the price dynamics implied by the model. For instance, in the context of the original Black-Scholes model, this would suggest that the price of the stock be taken as given and its price be determined as the present value of the expected stock price in the future under the risk-neutral distribution. However, in the context of term structure models, there is more than one underlying asset involved, since the prices of interest contingent claims depend, in principle, on the current prices of bonds of all maturities along the term structure. Furthermore, these bonds themselves have to be priced relative to the future possible movements in an arbitrage-free manner. Since there are several bonds involved, this translates into two types of no-arbitrage conditions. First, each bond has to be priced correctly with respect to its future price evolution. Second, the price movements across two or more bonds should not imply arbitrage opportunities. The latter feature is indeed the key distinguishing feature of term-structure models. To illustrate, the pricing of the stock options involves a single-step exercise, given the stochastic process for stock prices. In contrast, the pricing of bond options is, in principle, a two-step effort requiring first the derivation of the bond price process from the interest-rate process and then the pricing of bond options as a second step.

An additional issue that arises in specifying the current term structure (the underlying assets) is whether to specify it in terms of spot interest rates or forward interest rates. Although at a broad conceptual level the two specifications would be equivalent, the input data and the details of the models may be quite different, depending on the stochastic processes assumed. As an example, the derivation of results on the equivalent martingale measure may be easier with the forward-rate specification. Also, as is discussed below, it may be in line with no-arbitrage restrictions to assume that the volatility of forward rates is constant, while this may be difficult to do with spot interest rates. Thus, for particular purposes, one representation may turn out to be more convenient than the other.

The specification of the interest-rate process, either directly assumed or derived from the underlying state variables, has to be in line with certain stylized facts. Although some of these stylized facts can be discussed at a broad conceptual level, others can be related only to the empirical evidence. The basic issue that separates models is the number of factors used to describe the term structure movements over time. The simplest structure would involve only one factor so that all bond returns are instantaneously perfectly correlated, with the single factor usually being taken as the short-term interest rate. Generalization to two or more factors requires care on two grounds. First, from a conceptual perspective, the relationship between the factors has to be explicitly taken into account, in order to be consistent with the no-arbitrage requirement. For example, if the two factors are a short-term interest rate and a long-term interest rate, it should be borne in mind that the long-term bond is related to a series of short-term bonds in a risk-neutral world and hence the movements have to be tied together in some fashion. As a practical matter, due to problems of computational complexity, the number of factors is usually restricted to a maximum of two.[2]

From an empirical viewpoint, there is considerable latitude in the choice of the two factors. Examples include the short-term interest rate and the long-term interest rate (e.g., Brennan and Schwartz [1979]), the short-term interest rate and a spread between the long-term and short-term interest rates (Schaefer and Schwartz [1984]), the short-term interest rate and the rate of inflation (Cox, Ingersoll, and Ross [1985b]), the short-term interest rate and its volatility (Longstaff and Schwartz [1992]), the short rate and duration (Schaefer and Schwartz [1984]), and the short rate and its mean (Das and Foresi [1996]). At a more detailed level, there is the question of what the best proxies are for each of these variables. For instance, should the short-term interest rate be a three-month rate or a four-year rate?[3] The empirical analysis could be at two levels, either to see how various alternative specifications fit the term structure data or to examine how the contingent claims models implied by these structures fit the prices of interest rate options such as caps, floors, and swap options.

The other aspects of the term structure specification are the distributional assumptions built in. Three issues are relevant for discussion here. First, is the process a pure diffusion, or does it allow for jumps? Second, in the case of diffusion process, are there any restrictions placed on the drift term, such as mean-reversion or some type of time-dependence, for example? Third, what is the specification of the diffusion term, and what implication does this have for the distribution of interest rates?

Turning to the first issue of jumps, most option pricing models assume diffusion processes for the price of the underlying asset. This is natural, given the dynamic hedging arguments on which contingent claim pricing models are based. However, empirical work has pointed to the existence of jumps in the interest-rate processes. Discontinuities in the price paths of the underlying asset can be accommodated in these models only with the imposition of risk-neutrality or the assumption of a state variable that correlates perfectly with the jump component.

The second issue of mean-reversion is an aspect of the process governing movements in the short-term interest rates that has often been postulated. Although the empirical evidence in support of the existence of mean reversion is somewhat weak, both academics and practitioners feel that this is an important stylized fact to be built into interest-rate option pricing models, particularly when applied to high-frequency data.

The third issue of the choice of the specification of the diffusion term is dictated by certain stylized facts tempered by analytical tractability. At a conceptual level, the main constraints imposed by these stylized facts would be the restriction that *nominal* interest rates stay *nonnegative*, with the mean-reversion term being taken into account. At an empirical level, the Gaussian and the lognormal distributions for interest rates are taken to be the polar cases, with other cases in between being characterized by the elasticity of the volatility in the diffusion term. Although the empirical evidence is so far inconclusive, it appears that the elasticity lies between one-half and unity (if stochastic volatility is not assumed). If the criterion of analytical tractability is used to obtain closed-form solutions for bond and contingent claim prices, the candidates for the interest-rate processes are three special cases: Gaussian and the square root processes (for which the elasticity parameter is one-half) and the lognormal case under some special circumstances.

3. An overview of the major approaches to term structure modeling

Although the literature in the area is vast, it turns out that five papers capture most of the considerations discussed in the previous section. Much of the other work in the area may be considered to be extensions or elaborations of these five papers. It would be useful, in a survey of this nature, to isolate the main features of these five approaches and the strengths and weaknesses based on the specific aspects of each of the various approaches, in order to provide a synthesis of the area and point out the lacunae in the present paradigm.

3.1. The Black-Scholes-Merton model

The Black-Scholes [1973] and Merton [1973b] papers establish the basic paradigm for the pricing of contingent claims. Although these papers primarily discuss the pricing of options on stocks, the application of the approach to the case of bond options is also discussed by Merton. The Merton modification of the basic Black-Scholes model views the stochastic process for the price of a long-term bond as being similar to that of a stock—that is, as a lognormal diffusion process,

$$dB/B = \mu dt + \sigma dz, \tag{1}$$

where B is the discount bond price, μ is the drift term or instantaneous expected return, σ is the volatility, and z is the standard Brownian motion.

The price of any contingent claim based on such a bond can be determined by constructing a dynamic hedge based on the underlying bond and the (short-term) riskless asset. The price of the contingent claim, determined by this replication argument, depends on the payoffs on this claim and can be determined by a variety of methods. The original method proposed by Black-Scholes involves solving the fundamental partial differential equation for the price of a contingent claim, subject to the boundary conditions defined by the payoffs of the claim. The alternative martingale approach, which is now more commonly used, involves the evaluation of the discounted expected values of the payoff option. For example, for a European call option on a bond, with maturity T, the modified Black-Scholes model proposed by Merton, implies that the price would be given by

$$C = BN(d_1) - Ke^{-r(T-t)}N(d_2), \tag{2}$$

where

$$d_1 = \frac{\ln(B/K) + (r + \sigma^2/2)(T - t)}{\sigma\sqrt{T - t}}, \qquad d_2 = d_1 - \sigma\sqrt{T - t},$$

where C is the price of the European call option, K is its exercise price, r is the instantaneous risk-free rate (assumed constant), σ is the volatility of the instantaneous change in the (log) bond price, and $T - t$ is the time to expiration. $N(\cdot)$ is the cumulative normal density function.

The above model is a fairly straightforward extension of the Black-Scholes model to the case of bond options and is used widely for relatively short-dated options on long-term bonds. The main virtue of this basic model for pricing bond options is its simplicity and intuitive appeal, given the widespread use of the Black-Scholes model. However, at a conceptual level, the model has some fairly serious shortcomings. The basic insight of the Black-Scholes approach is the risk-neutral valuation argument that the instantaneous drift term of the return on the underlying asset (which is preference-dependent and not observed) can be replaced by the short-term interest rate, r, which is observable. This does not raise any conceptual problems in the case of options on stocks, since it is possible, in principle, for stock prices to be volatile, while the risk-free rate is deterministic and, in particular, constant.[4] However, when applied to bond options, this argument is problematic. It is hard to imagine how the short-term interest rate can be nonstochastic while the long-term bond prices are volatile. More fundamentally, the no-arbitrage conditions between bonds of different maturities are not explored. Furthermore, the information contained the current term structure, which would be relevant to the characterization of the price of the underlying asset, the long-term bond, is not taken into account.

A second issue, that is probably more important for the valuation of bond options, is the specification of the volatility parameter. For the case of options on stocks, which are analyzed in the basic Black-Scholes framework, it is not unreasonable to suppose that the volatility of the return on the underlying asset is constant through time, since the asset is infinite lived. Thus, for stocks, it makes sense to assume that the range of possible stock prices changes gets larger as the expiration date of the option is moved further out in time. For the case of bonds, however, this is not a sensible assumption, since the price of a (default-free) bond on its maturity date is known with certainty. Hence, unlike the prices of stocks, bond prices must have an increasing range of values for a period but thereafter have a decreasing range, converging to a fixed price on the maturity date of the bond.[5]

In order to address these concerns, some researchers model the behavior of the volatility of the bond price over time, in an exogenous fashion, while preserving the basic Black-Scholes structure. These approaches include the Brownian-bridge process proposed by Ball and Torous [1983], for which the bond prices at the current date and the maturity date are "tied down," leaving the process in-between to fluctuate, similar to a suspension bridge. It is necessary therefore to specify the process followed by the bond price between the current date and the maturity date. Particular specifications of this process include the Macaulay duration-based specification of the volatility parameter suggested by Schaefer and Schwartz [1984]. The intuition behind this approach is that the duration of the bond is highest at the current date, declining to zero on the maturity date of the bond, so that the volatility parameter exhibits a similar behavior. Schaefer and Schwartz suggest that a time-varying parameter could then be built into the Black-Scholes framework to obtain at least a partial resolution to the volatility problem. However, in these ad hoc approaches, basic arbitrage conditions may be violated, and hence the solutions may be unsatisfactory.

A third problem with applying the Black-Scholes model to bond options is the implication for interest rates. Since the price of the bond follows a lognormal diffusion process, the instantaneous returns would be normally distributed. In a one-factor model, this would imply that the short-term interest rate is also normally distributed, which would mean that interest rates could become negative. An alternative to preserve the property of positive interest rates is to assume that interest rates are lognormally distributed. The Black-Scholes model can be modified for options on forward contracts to yield the Black [1976] model, which can be written as follows:

$$C_f = e^{-r(T-t)}[FN(d_1) - KN(d_2)]$$

$$d_1 = \frac{\ln(F/K) + \sigma^2/2(T-t)}{\sigma\sqrt{T-t}}$$

$$d_2 = d_1 - \sigma\sqrt{T-t},$$

where C_f is the price of a call option on a forward contract on the underlying asset, F is the forward price, and the other variables are as defined earlier. The d_1 and d_2 terms do not have the interest rate in them since it is already reflected in the forward price. The Black model is widely used to value interest-rate caplets and floorlets, which can be regarded roughly as options on the forward interest rate multiplied by a face value. In other words, the forward price and the strike price are replaced by the forward interest rate and the strike rate of the caplet or floorlet. Again, this solution may be practically quite useful, but is not properly grounded in an arbitrage-free specification.

3.2. The Vasicek model

Soon after the development of the Black-Scholes model, Vasicek [1977] developed a partial equilibrium model of the term structure of interest rates. Although the model does not fully use the dynamic hedging arguments of Black-Scholes, since it does not take the current term structure as given, it applies the continuous-time methodology and the concept of arbitrage that is central to the original Black-Scholes formulation.

At a general level, the model postulates that the short-term interest rate, the only state-variable in the model, follows a diffusion process

$$dr = \alpha(r, t)dt + \sigma_r(r, t)dz, \qquad (3)$$

where $\alpha(r, t)$ and $\sigma_r(r, t)$ are the instantaneous drift and standard deviation respectively of the process for $r(t)$ and z is a Brownian motion. Since r is the only state variable in the model (a key assumption of the model), the price of any discount bond, whose price depends on the current date, the maturity date, and the state variable, can be written by Ito's lemma as a function of these variables:

$$dB = B\mu(t)dt - B\sigma(t)dz. \tag{4}$$

Since there is only one stochastic variable driving all bond prices, it is possible to set up a hedge portfolio, consisting of bonds of two different maturities, that is instantaneously riskless. Note that this explicitly rules out, by assumption, other possible influences on the bond price process.

The arbitrage mechanism across bonds, which is similar in spirit to the Black-Scholes argument, yields the condition that the ratio of excess return to risk, the measure of the market price of risk, has to be equated across bonds:

$$\lambda(t, r) = \frac{\mu(t, T_1) - r(t)}{\sigma(t, T_1)} = \frac{\mu(t, T_2) - r(t)}{\sigma(t, T_2)}, \tag{5}$$

where T_1 and T_2 are the maturity dates of the two discount bonds and λ is the market price of risk. The arbitrage condition also yields a partial differential equation similar to the one derived by Black-Scholes, which can be solved subject to the boundary condition that the bond price at maturity has to be equal to its face value of unity:

$$\frac{\partial B}{\partial t} + (\mu + \lambda\sigma)\frac{\partial B}{\partial r} + \frac{1}{2}\sigma^2\frac{\partial^2 B}{\partial r^2} - rB = 0. \tag{6}$$

There is one important difference between the above equation and its Black-Scholes counterpart: the presence of the market price of risk, λ. Since this parameter would depend on the risk attitudes of investors, the model is not preference-free, unlike the Black-Scholes formulation. The reason this is so is not hard to seek: the Black-Scholes model takes the stock price, which reflects the market price of risk, as given. In contrast, the Vasicek model solves for all bond prices relative to each other. The only way to tie down the prices is by invoking the exogenous parameter, the market price of risk.

After deriving the general model, which cannot yield bond prices in closed form, Vasicek provides an explicit solution for the case of the Ornstein-Uhlenbeck (O-U) process for the case of a constant λ:

$$dr = \kappa(\ell - r)dt + \sigma dz, \tag{7}$$

where $\kappa > 0$ is the "elasticity" of the mean-reversion term and ℓ is the long-term mean toward which the short rate is pulled. The attractive feature of this process is that the process converges to a stationary distribution, unlike a diffusion process without the mean-reversion term. Furthermore, the process is in line with empirical work on interest rates, which suggests mean reversion in the short rate, although the evidence in support is somewhat mixed. Since the conditional expectation and variance of the process can be computed, it is possible to derive closed-form expressions for bond prices and interest rates, either spot or forward. However, all these expressions are dependent on the exogenous parameter, the market price of risk, λ, which is assumed to be constant. The latter assumption obviously implies strong restrictions on the preferences of investors.

The strength of the Vasicek formulation is that it focuses attention on the notion of arbitrage across bonds, in the spirit of Black-Scholes and Merton ([1971], [1973a], and [1973b]). It proposes a special case of a mean-reverting Gaussian interest-rate process that has the twin virtues of intuitive appeal and analytical tractability. The intuitive appeal rests on the concept of mean reversion that is reflected in this process. The closed-form expressions that the model yields allow for empirical testing without too much computational effort.

The main weaknesses of the model are the assumptions of an exogenous market price of risk, which means that the model is not preference free. Worse, particular choices of the parameter may imply arbitrage opportunities, causing the model to be internally inconsistent. Also, the model does not incorporate the observed current term structure, which means that the model prices do not reflect current market information (although this could be achieved with a choice of the market price of risk, λ, that varies with maturity). In turn, this means that a pricing of contingent claims would be subject to two possible sources of error: the error in explaining the current term structure and the errors in the model for pricing contingent claims, given the model term structure. The specific mean-reverting Gaussian model assumes a constant market price of risk, which may be inconsistent with the absence of arbitrage. Also, given that interest rates are normally distributed, there is a positive probability of negative interest rates, which is not reasonable.

The Vasicek model has been extended in three directions in subsequent research. First, alternative processes have been proposed for the stochastic process for the short-term interest rate [1977]. The work of Dothan [1978], Courtadon [1982], Cox, Ingersoll, and Ross [1985b], and Stapleton and Subrahmanyam [1993] can also be placed in this category, although the Cox-Ingersoll-Ross paper has several other additional aspects of interest, which will be discussed below. A general formulation which allows for non-Gaussian interest rates would be

$$dr = \kappa(\ell - r)dt + \sigma_r r^{\beta} dz \tag{8}$$

where β is between 0 and 1. $\beta = 1$ would be the lognormal case, while $\beta = 1/2$ permits an analytical solution as shown by CIR. Second, the Vasicek model has been extended to include multiple factors, in particular two factors, as in the work of Brennan and Schwartz [1979] and Longstaff and Schwartz [1992], among others. There is considerable latitude in picking the second factor—for instance, the long-term interest rate, the spread between the short rate and the long-term rate, inflation, and volatility of the short rate itself. One common structure proposed by Brennan and Schwartz [1979] includes a short rate which is mean reverting to the long rate, ℓ, the second factor:

$$dr = \kappa(\ell - r)dt + \sigma_1 r dz_1$$

$$d\ell = \sigma_2 \ell dz_2$$

$$dz_1 dz_2 = \rho dt,$$

where σ_1 and σ_2 are standard deviations of the short- and long-rate process respectively and ρ is the correlation between them. In this model, the long rate is taken to be the yield on a perpetual bond, which can be observed (approximately). However, the market price of risk is still required to be specified in the pricing of all bonds, since the short-rate process is not hedgeable. The attractive feature of a two-factor model is that it can explain a greater variety of term structure movements over time. The disadvantages are the reliance on an exogenous parameter, the market price of risk, which is not observable and the cumbersome numerical implementation of the model, since closed-form solutions are often difficult to obtain. Also, even such a two-factor model is unable to accommodate all shapes of the term structure, particularly the inverted hump shape.

A third direction in which the Vasicek model has been extended is to derive the prices of contingent claims. Since all contingent claims have to satisfy the Vasicek partial differential equation but with different boundary conditions, one can solve for prices of interest-rate options. However, the solution will involve the market price of risk parameter. If one uses the bond price equation as well, in principle one can eliminate λ from the pricing equation for contingent claims. This is essentially the Gaussian interest-rate model of Jamshidian [1989], which leads to closed-form solutions for options on discount bond prices, with the important difference that the λ parameter is not eliminated.

3.3. The Cox-Ingersoll-Ross model

The Cox-Ingersoll-Ross ([1985a], [1985b]) is the first general equilibrium model of asset prices that can also be applied to model the term structure of interest rates. Although there is a considerable amount of notation and technical detail in the general version of the model [1985a], the basic economic arguments can be summarized in an intuitive manner. The model assumes a single-good, continuous-time economy with n linear production activities, whose expected-return vector and variance-covariance matrix are specified. There is also a vector of state variables whose expected return and variance-covariance matrix are given. There are many identical consumer-investors who maximize a time-additive utility function by choosing their consumption and portfolio weights for each activity, as well as their holdings of contingent claims based on these activities. In equilibrium, the wealth is fully invested in all the production activities. The solution yields the optimal decision of the consumer-investor and the equilibrium interest rate as a function of the investor preferences, wealth, and attributes of the state variables. In addition, the model also yields the partial differential equation to be satisfied by all contingent claims.

The general specification of the model does not permit an explicit solution of the interest-rate process in the pricing model for bond prices or contingent claims. The particular specification used by CIR to yield such results involves two explicit assumptions: the preferences of investors are logarithmic, and the single-state variable describing the economy follows a square-root process. The motivation behind the first assumption is that portfolio choices are independent of the investors' wealth—that is, investors' decisions are myopic. This considerably simplifies the computation of the equilibrium, since it separates the present outcome of wealth from the future portfolio decisions. The second assumption permits the process for the short-term interest rate to be written as

$$dr = \kappa(\ell - r)dt + \sigma\sqrt{r}dz, \tag{9}$$

which is a particular case of the model in equation (8), with $\beta = 1/2$.

The CIR model has the attractive feature of being firmly grounded in a general equilibrium framework and hence is completely internally consistent. It makes explicit the variables on which the equilibrium bond prices depend and also derives the bond price dynamics. In contrast, partial equilibrium models make implicit assumptions about these variables. Apart from this general benefit, it avoids the problems associated with ad hoc specifications—for instance, for the market price of risk. CIR show, for instance, that assuming a simple linear structure for λ, the market price of risk, to be a function of r could lead to internal inconsistency where arbitrage is possible. Thus, there is a distinct advantage to determining the market price of risk endogenously. Turning to the specification of the short-rate process, it has several advantages, apart from analytical tractability and the existence of a steady-state distribution for the interest rate. The interest rates implied are in line both with intuition and the empirical findings. Under the square root process, at least for a range of parameter values, interest rates can never become negative or become absorbed at zero.[6] Furthermore, the process has the property that the variance of the interest rate increases with its level, a feature that accords with the stylized facts from the empirical work.

The general version of the CIR model provides a complete solution to the problem of equilibrium interest rates and bond prices. This model has no explicit solution, unless one specifies the model further. However, the problem with the specific version of the CIR model, which is implementable, is that it rests on the fairly specific assumptions of a square-root technology and logarithmic utility. The latter assumption implies that portfolio decisions in succeeding periods are independent of wealth at the end of the current period. Hence, investor preferences have a myopic property, which probably restricts the movements in the term structure over time. More general preferences, which allow dependence between the investors decisions over time, may allow for a wider range of possible term structure movements. However, this may prove to be difficult to solve analytically. In addition, as in the case of the Vasicek model, the CIR model does not take into account the current term structure observed in the market. Hence, since the model would not explain the present term structure perfectly, there would be arbitrage possibilities between the term structure implied by the model and that observed in the market. This mispricing by the model also would be reflected in the pricing of contingent claims, with the result that basic arbitrage relationships between the model prices for such claims and the observed term structure, such as put-call parity, would fail. This would severely restrict the applicability of the model to practical problems of valuation. In addition, the model would predict movements in the term structure that would systematically deviate from observed data. These drawbacks of the CIR approach, which would apply to any model that prices both bonds and interest-rate contingent claims, are mainly due to the ambitious nature of the task undertaken. Despite these problems, the model remains the only fully consistent equilibrium model of the term structure.

The one-factor version of the CIR model has been extended in several different ways. One extension is to explicitly incorporate the effect of expected inflation into the model. By interpreting the equilibrium result in the basic model in the context of a single-good economy, this approach views the square-root process as a more appropriate model for real rather than nominal interest rates. Gibbons and Ramaswamy [1993] and Brown and Schaefer [1994] use this interpretation in their empirical work and achieve some success in explaining real bond prices. To price nominal bonds, however, it is necessary to specify the inflation process or more generally the effect of inflation on the real economy. In their specific model, CIR [1985b] proposed a partial equilibrium approach to accommodate the uncertainty of inflation within bond pricing models by assuming that the expected inflation rates govern both the drift and the diffusion of the price level process. Another assumption in CIR [1985b] is that the state variable governing the production is not correlated with innovations to inflation. Thus, the treatment of inflation in CIR requires not only the exogenous determination of the rate of inflation in the economy but also the assumption of neutrality of money.

Pearson and Sun [1994] extend the original CIR specification for nominal bond pricing model by adding a constant term to the inflation and real interest rates. Their specification, which is the most general version of the "affine" yield model, assumes that the variance of the state variable is linear in the level of the state variable rather than strictly proportional to it. As discussed previously, the non negativity of the square-root process is appealing for nominal interest rates but may be a rather severe restriction for real interest rates. Pearson and Sun [1994] take this into account by introducing a constant term that changes the level of the boundary for the process, so that negative real interest rates can be accommodated. In order to test the assumption of neutrality of money, Sun [1992] allows for correlation between the state variable in the consumption growth and the innovation in inflation. In Pennacchi [1991], the expected inflation is modeled to govern production process by affecting the correlation as well as the expected return of real investments. The empirical results of these papers indicate that it may be unrealistic to assume that inflation is independent of the real economy. Given that the inflation process is exogenously specified in these models, another extension is to admit money into a more general equilibrium setting where the inflation is endogenously determined. For example, in Giovannini and Labadie [1991], the term structure of indexed bonds and nominal bonds is analyzed in the cash-in-advance model of Svensson [1985].

Longstaff and Schwartz [1992] and Chen and Scott [1992] extend the one-factor CIR model by including additional factors which follow the square-root process. These models, as well as the one by Duffie and Kan [1993], can be classified as multifactor affine models since these models are mathematically isomorphic and share the property that the short-rate bond yields and forward rates are affine in the underlying state variables. A remaining issue in these multifactor models is how to identify the observable factors in the implementation of the models. In fact, one of the biggest advantages of one-factor models is that the short rate provides sufficient information to the state of economy, so that the factor can be observed without any errors. In the Duffie and Kan [1993] two-factor model, factors can be identified by any two bonds yields. In Ho, Stapleton, and Subrahmanyam [1995], two factors are identified by the short rate and a forward rate. Longstaff and Schwartz [1992] observe that the conditional variance of the short-term interest rate is also a linear function of state variables, so that state variables can be identified by the short rate and the conditional volatility. Because conditional volatility can be measured with satisfactory precision due to the progress in econometric technique related to the GARCH type model, the Longstaff and Schwartz model can be implemented without losing too much of the information embedded in the yield curves. Given that there is no consensus regarding the choice of identifying state variables in multifactor models, the only alternative is to compare the pricing errors for different choices of the two factors. Although there is some work in this direction, more research remains to be done using alternative data sets and econometric methodologies.

3.4. The Heath-Jarrow-Morton model

Although not an equilibrium model, the Heath-Jarrow-Morton [1992] model is a complete model of the term structure specified in an arbitrage-free manner in the spirit of Harrison and Kreps [1979] and Harrison and Pliska [1981]. The essential feature of the approach is its characterization of the sufficient conditions under which a unique equivalent martingale measure exists. Given such conditions, markets are complete, and the valuation of contingent claims is well-defined at a conceptual level, although issues of implementation remain.[7] At one level, the HJM approach can be viewed as a multifactor extension of the basic Black-Scholes framework, if one could imagine a multiasset economy where forward prices of assets follow general diffusion processes. Under certain conditions for the drift terms and volatilities, an equivalent martingale measure exists, and hence all claims, defined in a completely general manner based on the multiple assets can be valued by taking the expectation of their payoffs under the equivalent martingale measure. As in the case of the Black-Scholes framework, the current prices of the underlying assets—in this case, the entire term structure today—have to fit the model perfectly. The only additional feature of the term structure problem, as compared to the general problem of asset pricing, is that there are restrictions on the pricing of bonds of different maturities. Since the bonds of different maturities are related in a specific manner, unlike assets in general, one has to be careful to specify the price process. Specifically, there should be no arbitrage possibilities across bonds, in the sense that the instantaneous return on one bond should not stochastically dominate that of another. Thus, in a binomial model, for example, the return in the first period on a two-period bond should straddle the return over the same period from a one-period bond: in one state in the future it would be higher, and in the other it would be lower.

The basic set-up of the model is similar in spirit to the Vasicek model with the important difference that we are dealing with forward rather than spot interest rates. The process for the forward interest rates is written as

$$df(t, T) = a(t, T)dt + b(t, T)dz, \tag{10}$$

where $a(t, T)$ and $b(t, T)$ are the drift and diffusion terms of the forward rate process, t is the current date, T is the maturity date, and z is a Brownian motion. $f(t, T)$ is the instantaneous forward interest rate at time t for delivery at date T. In order to keep the notation simple, we consider only one source of uncertainty. The spot interest rate process can be written as a special case of the above as

$$dr(t, t) = a(t, t)dt + b(t, t)dz, \tag{11}$$

where the second arguments of $a(\cdot)$ and $b(\cdot)$ vary over time. Note that the specification of spot rate process here is the same as in the general version of the Vasicek model, except that in the latter case the forward rate process is not similarly defined. Thus, the Ornstein-Uhlenbeck version of the Vasicek model as well as the CIR model are special cases of the above formulation, with restrictions being placed on $a(t, t)$ and $b(t, t)$.

The discount bond price can be defined by integrating over the entire structure of forward rates. Also, since the forward rates can be written as

$$f(t, \tau) = f(0, \tau) + \int_0^t a(s, \tau)ds + \int_0^t b(s, \tau)dz_s, \tag{12}$$

the bond price process can be written as the definitional relationship

$$\begin{aligned} \ln B(t, T) &= -\int_t^T f(t, \tau)d\tau \\ &= -\int_t^T f(0, \tau)d\tau - \int_t^T \int_0^t a(s, \tau)dsd\tau - \int_t^T \int_0^t b(s, \tau)dz_s d\tau. \end{aligned} \tag{13}$$

Superficially, this step resembles the analysis in the Vasicek model, where the bond price is written by integrating over the path of spot interest rates. The important difference is that in the Vasicek case, the analysis is driven by the *assumption* of a one-factor model, where the factor is the short rate. In other words, there is no guarantee that some other state variable has not been left out. In the HJM case, the analogous condition for the bond price is in terms of forward rates and hence follows definitionally.

Another important difference between modeling the spot-rate process versus the forward-rate process stems from the structure of the instantaneous volatility parameter. Since the forward interest-rate process can be assumed to be trend stationary, it is possible to integrate over the $a(\cdot)$ and $b(\cdot)$ and write the process for the bond price as

$$dB(t, T) = B(t, T)[\mu(t, T)dt + \sigma(t, T)dz]. \tag{14}$$

This is the same as equation (14) in the Vasicek model, except that equation (14) is derived from the definition of bond price as a function of the forward interest rates, rather than by assuming a single-state variable model as Vasicek does.

At this stage, it is useful to point out the essential difference between the martingale approach of Harrison and Kreps [1979] and Harrison and Pliska [1981] and the Black-Scholes approach to the pricing of contingent claims. The Black-Scholes model starts with a specific structure and derives the necessary condition for the absence of arbitrage, which happens to yield a risk-neutral valuation relationship. In contrast, the martingale approach aims to derive the sufficient conditions under which the market price of risk parameters can be eliminated and contingent claims can be valued under the equivalent martingale measure.

The key result of the HJM is to show the equivalence between three alternative conditions that lead to preference-free pricing:

- The market price of risk is independent of the vector of horizon dates;
- A unique equivalent martingale measure exists; and
- There is a particular relationship between the drift and volatility terms of the forward-rate process and the market price of risk.

Operationally, the last condition is the one that can be applied directly to eliminate the market price of risk.

The main insight of the HJM method can be illustrated for the special case of a single-factor model. In this case, the last condition implies

$$a(t, T) = b(t, T)[\lambda(t) + \int_t^T b(t, \tau)d\tau]. \tag{15}$$

If we assume further that $b(t,\tau) = b > 0$—that is, that the forward-rate volatility is constant—equation (15) can be written as

$$a(t, T) = b\lambda(t) + b^2(T - t), \tag{16}$$

which on substituting into the forward-rate process, and using the transformation from the original Brownian motion to its risk-neutral counterpart

$$\hat{z} = z + \int_0^t \lambda(s)ds \tag{17}$$

yields

$$f(t, T) = f(0, T) + b^2t(T - t/2) + b\hat{z}, \tag{18}$$

where the drift term has been eliminated and the Brownian motion is replaced by the risk-neutral counterpart $\hat{z}$.

From the above expression for the forward-rate process, it is easy to write down the spot rate process as a special case. Also, the bond price can be written down by integration over the forward rate between t and T, in terms of the risk-neutral process. It is straightforward now to value any contingent claim by computing its payoff, evaluating its expectation under the risk-neutral distribution, and discounting it to present using the integral over the instantaneous forward rates over the life of the contingent claim. It should also be emphasized that all preference-dependent parameters have been eliminated from the valuation model.

3.5. The Ho-Lee model

The Ho-Lee [1986] model deserves mention as a special case of the HJM model because it is the first approach to the implementation of arbitrage-free interest-rate processes. In continuous-form, it is essentially the single-factor example analyzed in the previous section. Indeed, without mean reversion, it is simply a one-factor version of the HJM model where the forward rates follow arithmetic Brownian motions with constant volatility coefficients on the diffusion term. However, since this model was originally formulated as a discrete-time model, it is also useful to think of it as an extension of the Cox-Ross-Rubinstein [1979] (CRR) model. The basic structure of the model can also be stated as a special case of the Vasicek model without mean-reversion

$$dr = \theta(t)dt + \sigma(t)dz. \tag{19}$$

The distinguishing feature of the model is its practical appeal due to its emphasis on implementation, although the HJM model provides the conceptual underpinnings of the Ho-Lee formulation. Also, the Ho-Lee model has the advantage of being an intuitively appealing special case of the HJM model.

The Ho-Lee approach involves fitting a binomial lattice for discount bond prices, taking the current term structure as given and imposing the restriction that the bond price is "pulled to par" at maturity. Since the model assumes only one factor, the movements in the entire discount function over time can be explained by just one state variable. The evolution of this state variable is captured by what Ho-Lee call a perturbation function, which simply represents the up and down movements in the binomial tree, which is time-dependent, unlike the basic Cox-Ross-Rubinstein formulation. These perturbation functions are specified as deviations from the implied forward price function. Hence the risk-neutral probability has to be specified such that the expected value across the two states adds up to unity, since the analysis is in terms of forward prices. This is strictly analogous to the approach of Cox, Ross, and Rubinstein with three minor differences. First, the analysis is in terms of forward prices rather than spot prices. Second, the up and down movements are time-dependent. Finally, in the Ho-Lee model, the whole term structure, rather than a single asset price, is shifted up or down.

The key attribute of the Ho-Lee model is that, by construction, there is no arbitrage allowed between the pricing along the lattice that is constructed and the current market interest rates. However, since no equilibrium model is specified, there is no question of explaining the current term structure, either. Further, once the lattice is constructed and the rates change, a gap will open up between the actual rates observed and the pricing along the lattice, one time-step later. This problem of time-inconsistency is faced by all current arbitrage-based models of the term structure.

Several extensions and modifications of the Ho-Lee approach have been proposed, prominent among them being the models of Black, Derman and Toy [1990] and Hull and White [1990]. In these alternative models, there are differences in the specification of the interest rate process in terms of variables such as mean-reversion, the time-dependence of the volatility and drift terms, and the elasticity of the diffusion. In addition, there are details relating to implementation, such as whether the lattice is binomial or trinomial, and whether it is constructed for spot or forward interest rates.

4. Empirical issues

The empirical examination of term structure models probably calls for a separate survey, focusing on the econometric issues. However, it would be useful to review briefly the main issues that have been addressed in the empirical work to date and what the findings are, so far. Models of the term structure can be tested either directly or in terms of their implications for the pricing of interest-rate contingent claims. The direct test of the term structure models can be conducted either cross-sectionally, across bonds of different maturities, or with time-series observations.

The major advantage of cross-sectional tests is that they explicitly identify the market price of risk, which is fundamental to the modeling of the term structure, from a theoretical perspective. The drawback, however, is that it is difficult to test a whole class of models in a nested fashion, so that models can be rejected or not at various levels. In contrast, time-series tests cannot directly impose the restriction on the market price of risk across bonds without extremely cumbersome computation but can test the models in a nested fashion. Thus, the time series tests are tests of "goodness of fit" to the stochastic process for interest rates, while the cross-sectional tests are tests of specific bond pricing models, and they nest a test of the process by construction.

Brown and Dybvig [1986] were the first to test the CIR model with cross-sectional data.Their broad conclusion is that the estimates of the long-term mean, to which the short rate reverts, as well as the instantaneous volatility parameter are unstable over time. Thus, the special form of the basic CIR model may be misspecified, although an extended version, along the lines discussed above, may better explain the bond price dynamics. However, this extended specification may have problems of under identification of the parameters. Gibbons and Ramaswamy [1993] use generalized method of moments (GMM) procedures to test the unconditional distribution of the steady-state interest-rate process and find that the CIR model cannot be rejected in cross-sectional data. However, this failure to reject may stem from the weak power of their test. In contrast, Pearson and Sun [1994] used the more powerful maximum likelihood estimate (MLE) procedure and were able to reject the CIR model in nested tests.

The time-series tests of term structure models infer the parameters from short-term interest rates (typically one-month Treasury bill rates) in tests of the underlying stochastic process. Apart from the difficulty in identifying the market price of risk, the tests require much longer data series, since cross-sectional data are not used. Also, since the data are obtained from discrete sampling, it is necessary to use the Euler approximation to discretize the continuous-time process. On the other hand, for data of frequencies up to monthly, there is little difference between using an exact estimator and the Euler approximation.[8]

Marsh and Rosenfeld [1983] and Chan, Karolyi, Longstaff, and Sanders (CKLS) [1992] use more general processes to nest the CIR model. They both reject the CIR model, and in the case of CKLS, they find that the elasticity term is about 1.5! This would imply that the short-term interest rate process is nonstationary, which is out of line with any theory of the term structure. Recent work by Eom [1994] indicates that conclusions of these tests may be due to inefficient estimation methods.

A test of the contingent claims pricing implications of alternative term structure models was first conducted in a comprehensive manner by Amin and Morton [1994]. Their results indicate that the implied volatilities from the six alternative special cases of the HJM formulation are unstable. Although the goodness of fit improves with two-parameter specifications, the instability gets worse. The empirical evidence to date suggests that one-factor models that are interesting special cases of HJM may be misspecified because the parameter estimates are unstable. However, Amin and Morton show that estimates from a one-parameter volatility model outperform those from a two-parameter volatility model and that the Gaussian models do better than lognormal ones! The one-parameter model does better possibly because it is more robust over time and is less susceptible to the time inconsistency issue.

There are two possible directions for new empirical research on the term structure. The first would be to look at multifactor models more closely, along the lines of Longstaff and Schwartz [1992]. The second would be to specify the tradeoff between fitting the current term structure perfectly in cross-sectional data versus forecasting the future term structure with time-series data. Of course, fitting better cross-sectionally means that time inconsistency may be exacerbated. Furthermore, it is necessary to apply the models to data on important interest-rate derivatives such as caps, floors, and swaptions.

5. A brief synthesis of models of the term structure

We are now in a position to put together a broad picture of the field emphasizing the main areas where we have fairly good answers to the issues raised. In doing so, we use as a basis the papers surveyed in Section 3 but supplemented by some others that focus on issues of implementation.

At a general level, following CIR, we have a formal understanding of connection between technology, preferences, and asset prices in an equilibrium setting, with particular reference to interest rates. At a general equilibrium level, we are aware of a set of sufficient conditions that yield closed-form solutions for the short-term interest rate and for bond prices: square root technology and logarithmic utility functions. Two other combinations of these variables can also be solved in a similar fashion—lognormal diffusions and constant proportional risk aversion (CPRA) or normal diffusion and constant absolute risk aversion (CARA). However, the bond prices and interest-rate processes they yield have undesirable properties such as negative interest rates or even bond prices.

Although it is important to ensure that the general equilibrium foundations of the models in the area are well established, it is not necessary to go back to them in every model that is developed. At the next level, we could work with the pure-exchange partial equilibrium models of the type developed by Merton [1973a] and several others. Special cases of these models, such as the Gaussian interest-rate model of Vasicek, in a partial equilibrium setting, are well understood. This approach provides a necessary condition for the absence of arbitrage, which essentially amounts to equating the market price of risk implied across bonds. This restriction leads to a partial differential equation that has to be satisfied by all bonds and contingent claims, which can be solved subject to their respective boundary conditions. Both the general and partial equilibrium approaches have the disadvantage of not perfectly fitting the current term structure; consequently, standard arbitrage relationships such as put-call parity would fail with respect to current market prices. Further, these errors could get further compounded in the pricing of contingent claims in these models.

The next step of establishing a sufficient condition for arbitrage-free pricing is taken in the approach of HJM. The important feature of this approach is that the market price of risk parameters are eliminated by working with the equivalent martingale measures. Furthermore, these models are fully consistent with the current observed term structure. The major difference between the equilibrium and the martingale approaches is that in the former, we are forced to derive the market price of risk explicitly or specify it exogenously. In the latter approach, we take the forward rate processes as given and hence are able to eliminate the market price of risk parameter altogether. The problem is that in the equilibrium approach, the market price of risk can be derived explicitly only in special cases; on the other hand, specifying it exogenously may lead to problems of internal consistency. The martingale approach is perfectly adequate and, further, practically implementable, if one is concerned only with deriving prices of bonds and interest-rate contingent claims rather than with specifying the equilibrium supporting these prices.

Special cases of the HJM set up have been derived, which are implementable. Most of these use a single-factor structure in the spirit of Ho-Lee. Implementation of binomial lattices in various cases including normal and lognormal interest rates, different structures for mean reversion, time-dependent volatility, and so on has led to a fairly detailed understanding of the biases from the various alternative specification. In the special case of Gaussian model, closed-form solutions can be derived as shown by Jamshidian [1989]. Although the HJM approach is a multifactor approach, implementation of even a two-factor model has been quite cumbersome computationally, although there has been some recent work in the area proposing improvements in efficiency.

6. Conclusions and future directions

It is fair to conclude, based on the above survey, that we have progressed considerably in our understanding of term structure modeling and the pricing of interest-rate contingent claims. In particular, we are aware of the importance of building models that are arbitrage-free, in order to be internally consistent.

It is clear, however, that for practical purposes, it is simpler to use models such as Black-Scholes or Black, which can be stated in closed-form.[9] These models may work quite adequately even though they are not theoretically well grounded. For example, the Black-Scholes model is often employed for the valuation of short-term options on long-term bonds with satisfactory results. Similarly, for the valuation of interest-rate caps and floors, the widespread use of the Black model may be justified on practical grounds. However, for many other applications such as the valuation of path-dependent options on interest rates, or even American-style options, such as mortgage-backed securities, it is important to model the interest-rate process in an arbitrage-free manner. The methods discussed in the previous sections become very relevant for such purposes.

However, several gaps remain in our understanding. Although, the HJM model is well specified from an arbitrage perspective, it suffers from the problem of time inconsistency. In other words, the term structure has to be recalibrated every period to fit the market prices of bonds. Future recalibration is not taken into account in fitting the tree on any given date. It is not clear how this general problem can be solved, although it appears that some sort of Bayesian procedure is necessary.

At a different level, the equilibrium underpinnings of the HJM model are unclear. Without attempting to go to the full-blown general equilibrium level of CIR, which is intractable in its general form, the issue is whether it is possible to characterize the equilibrium on which the HJM specification rests. Specifically, what restrictions can be placed on the key parameter, the market price of risk, based on stylized facts, and how does this accord with the HJM model? Some work along these lines—for example, by Ritchken and Sankarasubramanian [1995b]—is under way. However, more research is required in this area.

Extensions of the basic model to include more general structure such as stochastic volatility and jumps in the interest-rate process have been undertaken—for example, the work of Ahn and Thompson [1988], Longstaff and Schwartz [1992], Das ([1994], [1995]) and Das and Foresi [1996]. However, these have to be related more closely to the general HJM framework.

At the level of implementation, single-factor versions of the HJM model with alternative specifications to incorporate stylized facts such as mean reversion, time-dependent, drift, and volatility terms, as well as alternative processes for the diffusion term, are now fairly well understood. However, it is necessary to explore further multifactor versions of the HJM model. In particular, since multifactor models are difficult to work with from a computational standpoint, more work is required to simplify the structure to make such models Markovian and hence implementable.[10]

At an empirical level, we need to gain a better understanding of the number of factors that would be required to explain term structure movements adequately. In addition, more work is required on assessing the importance of mean reversion and the elasticity terms for the short-rate process, as well as the role of jumps in interest rates.

Acknowledgments

I would like to thank Günter Franke, Jingzhi Huang, Richard Stapleton, and especially Sanjiv Das, Young-Ho Eom, and Rangarajan Sundaram for helpful discussions.

Notes

1. Recently, a few papers have claimed that an equilibrium model can be developed to support virtually any kind of interest-rate process. See Boldrin and Montrucchio [1986], for example. However, even if one accepts such a claim, it is necessary to develop an explicit model for several applications.
2. It is often convenient to depict term structure movements in terms of (1) parallel shifts, (2) twists, and (3) changes in curvature, which effectively implies a three-factor model. There are some three-factor models available such as Chen and Scott [1993], although the implementation of such models for general bond options would be extremely complex.
3. See Nelson and Schaefer [1983] and Elton, Gruber, and Michaely [1990], for example, for a detailed discussion of the considerations involved in making the choice about the number and type of factors and the empirical evidence from the United Kingdom and United States, respectively.
4. Merton [1973] also provides the solution for the value of an option when the interest rate is stochastic. In that case, Merton's model is similar to the basic Black-Scholes model, with an adjustment for the volatility of the log bond price.
5. However, it may be sensible to model bond options based on the *forward* price of bonds, using the Black-Scholes-Merton approach. This approach is widely used in practice to price European-style option products, such as interest-rate caps and floors, as discussed below.
6. The problem with absorbing and reflecting barriers at zero is that they admit arbitrage opportunities, since in both cases, the future direction of interest rates is known with certainty when the interest rate is at the barrier. Gaussian models do not have this problem, although they have the other problem of negative interest rates.
7. A parallel paper by Artzner and Delbaen [1989] also proposes a martingale approach to modeling the term structure. However, this approach suggests a two-step procedure to price contingent claims, first to price zero-coupon bonds and then to price contingent claims. The Heath-Jarrow-Morton procedure, in contrast, avoids this two-step procedure by taking the bond prices and forward price processes as given.
8. This further validates similar results Bergstrom [1972] demonstrated long ago.
9. A variation of these models, particularly for the pricing of interest-rate caps and floors, is the alternative approach suggested by Stapleton and Subrahmanyam [1993].
10. Some work in this direction has been done by Ritchken and Sankarasubramanian ([1995a], [1995b]) and Pennacchi, Ritchken, and Sankarasubramanian [1996].

References

AHN, C.M., and THOMPSON, H.E. [1988]: "Jump-Diffusion Process and Term Structure of Interest Rates," *Journal of Finance*, 43, 155–174.
AMIN, K.I., and MORTON, A. J. [1994]: "Implied Volatility Functions in Arbitrage-Free Term Structure Models," *Journal of Financial Economics*, 35, 141–180.
ARTZNER, P., and DELBAEN, F. [1989]: "Term Structure of Interest Rates: The Martingale Approach," *Advances in Applied Mathematics*, 10, 95–129.
BERGSTROM, A.R. [1972]: "The History of Continuous-Time Econometric Models," *Econometric Theory*, 4, 365–383.
BALL, C.A., and TOROUS, W.N. [1983]: "Bond Price Dynamics and Options," *Journal of Financial and Quantitative Analysis*, 18, 517–531.
BLACK, F. [1976]: "The Pricing of Commodity Contracts," *Journal of Financial Economics*, 3, 167–179.
BLACK, F., DERMAN, E., and TOY, W. [1990]: "A One-Factor Model of Interest Rates and Its Application to Treasury Bond Options," *Financial Analysts Journal*, 46, 33–39.
BLACK, F., and KARASINSKI, P. [1991]: "Bond and Option Pricing When Short Rates Are Lognormal," *Financial Analysts Journal*, 47, 52–59.
BLACK, F., and SCHOLES, M. [1973]: "The Pricing of Options and Corporate Liabilities," *Journal of Political Economy*, 81, 637–659.
BOLDRIN, M., and MONTRUCCHIO, L. [1986]: "On the Indeterminacy of Capital Accumulation Paths," *Journal of Economic Theory*, 40, 26–39.
BRENNAN, M.J., and SCHWARTZ, E.S. [1979]: "A Continuous Time Approach to the Pricing of Bonds," *Journal of Banking and Finance* 3, 133–155.
BROWN, S., and DYBVIG, P.H. [1986]: "The Empirical Implications of the Cox, Ingersoll, Ross Theory of the Term Structure of Interest Rates," *Journal of Finance*, 41, 616–628.
BROWN, R.H., and SCHAEFER, S.M. [1994]: "The Real Term Structure of Interest Rates and the Cox, Ingersoll and Ross Model," *Journal of Financial Economics*, 35, 3–42.
CHAN, K.C., KAROLYI, G.A., LONGSTAFF, F.A., and SANDERS, A.B. [1992]: "An Empirical Comparison of Alternative Models of the Short-Term Interest Rate," *Journal of Finance*, 47, 1209–1228.
CHEN, R.-R., and SCOTT, L. [1992]: "Pricing Interest Rate Options in a Two-Factor Cox-Ingersoll-Ross Model of the Term Structure," *Review of Financial Studies*, 5, 613–636.
CHEN, R.-R., and SCOTT, L. [1993]: "Maximum Likelihood Estimation for a Multifactor Equilibrium Model of the Term Structure of Interest Rates," *Journal of Fixed Income*, 3, 14–31.
COURTADON, G. [1982]: "The Pricing of Options on Default-Free Bonds," *Journal of Financial and Quantitative Analysis*, 17, 75–100.
COX, J.C., INGERSOLL, J.E., and ROSS, S.A. [1985a]: "An Intertemporal General Equilibrium Model of Asset Prices," *Econometrica*, 53, 363–384.
COX, J.C., INGERSOLL, J.E., and ROSS, S.A. [1985b]: "A Theory of the Term Structure of Interest Rates," *Econometrica*, 53, 385–407.
COX, J.C., ROSS, S.A., and RUBINSTEIN, M. [1979]: "Option Pricing: A Simplified Approach," *Journal of Financial Economics*, 7, 229–263.
DAS, S.R. [1994]: "Poisson-Gaussian Processes and the Bond Markets," Working Paper, Harvard Business School.
DAS, S.R. [1995]: "Exact Solutions for Bond and Option Prices When Interest Rates Follow Jump-Diffusion Process," Working Paper, Harvard Business School.
DAS, S.R., and FORESI, S. [1996]: "Exact Solutions for Bond and Option Prices with Systematic Jump Risk," *Review of Derivatives Research* 1, 7–24.
DOTHAN, L.U. [1978]: "On the Term Structure of Interest Rates," *Journal of Financial Economics* 6, 59–69.
DUFFIE, D., and KAN, R. [1993]: "A Yield-Factor Model of Interest Rates," Working Paper, Stanford University.
ELTON, E., GRUBER, M., and MICHAELY, R. [1990]: "The Structure of Spot Rates and Immunization," *Journal of Finance*, 45, 629–642.
EOM, Y.H. [1994]: "In Defense of the Cox, Ingersoll and Ross Model: Some Empirical Evidence," Working Paper, New York University.
GIBBONS, M.R., and RAMASWAMY, K. [1993]: "A Test of the Cox, Ingersoll, and Ross Model of the Term Structure," *Review of Financial Studies*, 6, 619–658.

GIOVANNINI, A., and LABADIE, P. [1991]: "Asset Prices and Interest Rates in Cash-in-Advance Models," *Journal of Political Economy*, 99, 1215–1251.

HARRISON, M., and KREPS, D. [1979]: "Martingales and Arbitrages in Multiperiod Security Markets," *Journal of Economic Theory*, 20, 381–408.

HARRISON, M., and PLISKA, S. [1981]: "Martingales and Stochastic Integrals in the Theory of Continuous Trading," *Stochastic Processes and Their Applications*, 11, 215–260.

HEATH, D., JARROW, R., and MORTON, A. [1992]: "Bond Pricing and the Term Structure of Interest Rates: A New Methodology," *Econometrica*, 60, 77–105.

HO, T.S.Y., and LEE, S.B. [1986]: "Term Structure Movements and Pricing Interest Rate Contingent Claims," *Journal of Finance*, 41, 1011–1029.

HO, T.S., STAPLETON, R., and SUBRAHMANYAM, M.G. [1995]: "A Two-Factor No-Arbitrage Model of the Term Structure of Interest Rates," Working Paper, Lancaster University.

HULL, J., and WHITE, A. [1990]: "Pricing Interest-Rate Derivative Securities," *Review of Financial Studies*, 3, 573–592.

HULL, J., and WHITE, A. [1993]: "One-Factor Interest Rate Models and the Valuation of Interest-Rate Derivative Securities," *Journal of Financial and Quantitative Analysis*, 28, 235–253.

HULL, J., and WHITE, A. [1994]: "Numerical Procedures for Implementing Term Structure Models II: Two-Factor Models," *Journal of Derivatives*, 2, 37–47.

JAMSHIDIAN, F. [1989]: "An Exact Bond Option Formula," *Journal of Finance*, 44, 205–209.

LONGSTAFF, F.A., and SCHWARTZ, E.S. [1992]: "Interest Rate Volatility and the Term Structure: A Two-Factor General Equilibrium Model," *Journal of Finance*, 47, 1259–1282.

MARSH, T.A., and ROSENFELD, E.R. [1983]: "Stochastic Processes for Interest Rates and Equilibrium Bond Prices," *Journal of Finance*, 38, 635–646.

MERTON, R.C. [1971]: "Optimum Consumption and Portfolio Rules in a Continuous Time Model," *Journal of Economic Theory*, 3, 373–413.

MERTON, R.C. [1973a]: "An Intertemporal Capital Asset Pricing Model," *Econometrica*, 41, 867–887.

MERTON, R.C. [1973b]: "The Theory of Rational Option Pricing," *Bell Journal of Economics and Management Science*, 4, 141–183.

NELSON, C.R., and SCHAEFER, S. [1983]: "The Dynamics of the Term Structure and Alternative Portfolio Immunization Strategies," in *Innovations in Bond Portfolio Management*, G.G. Kaufman, G.O. Bierwag, and A. Toevs (Eds.), JAI Press.

PENNACCHI, G.C. [1991]: "Identifying the Dynamics of Real Interest Rates and Inflation: Evidence Using Survey Data," *Review of Financial Studies*, 4, 53–86.

PENNACCHI, G.C., RITCHKEN, A.P., and SANKARASUBRAMANIAN, L. [1996]: "On Pricing Kernels and Finite State Variable Heath Jarrow and Morton Models," *Review of Derivatives Research*, 1, 87–99.

PEARSON, N.D., and SUN, T.S. [1994]: "An Empirical Examination of the Cox, Ingersoll, Ross Model of the Term Structure of Interest Rates," *Journal of Finance*, 49, 1279–1304.

RITCHKEN, A.P., and SANKARASUBRAMANIAN, L. [1995a]: "Lattice Models for Pricing Interest Rate Claims," *Journal of Finance*, 50, 719–737.

RITCHKEN, A.P., and SANKARASUBRAMANIAN, L. [1995b]: "Volatility Structures of Forward Rates and the Dynamics of the Term Structure," *Mathematical Finance*, 5, 55–72.

SCHAEFER, S.M., and SCHWARTZ, E.S. [1984]: "A Two-Factor Model of the Term Structure: An Approximate Solution," *Journal of Financial and Quantitative Analysis*, 19, 413–424.

STAPLETON, R.C., and SUBRAHMANYAM, M.G. [1993]: "Analysis and Valuation of Interest Rate Options," *Journal of Banking and Finance*, 17, 1079–1095.

SUN, T.S. [1992]: "Real and Nominal Interest Rates: A Discrete-Time Model and Its Continuous-Time Limit," *Review of Financial Studies*, 5, 581–611.

SVENSSON, L.E.O. [1985]: "Currency Prices, Terms of Trade, and Interest Rates: A General Equilibrium Asset-Pricing Cash-in-Advance Approach," *Journal of International Economics*, 18, 17–42.

VASICEK, O. [1977]: "An Equilibrium Characterization of the Term Structure," *Journal of Financial Economics*, 5, 177–188.

The Geneva Papers on Risk and Insurance Theory, 21: 29-33 (1996)

Some Remarks on Modeling the Term Structure of Interest Rates

GÜNTER FRANKE
Fakultät für Wirtschaftswissenschaften und Statistik, Universität Konstanz, D-78434 Konstanz, Germany

Marti Subrahmanyam [1996] provides an excellent review of models that deal with the dynamics of the term structure of interest rates. Most of this work has been accomplished in the last twenty years, partly stimulated by academic interest to apply the Black-Scholes framework to the pricing of interest-rate options but also by practitioners' needs to manage interest-rate risks. In the following, I shall first put the theoretical work on the term structure into the broader perspective of capital market research in order to highlight some important differences between valuation of stocks and of bonds. Second, I shall address some specific issues in modeling the term structure to reveal potential deficiencies of the current state of the art. Third, I shall discuss some implications for financial risk management.

1. Valuation of bonds versus valuation of stocks

Superficially speaking, we observe returns of bonds and of stocks so that we can easily derive average returns and return volatilities for both in a similar manner. But there is an important difference between bonds and stocks. We know the future payoffs of default-free, fixed-rate bonds, but we have enormous difficulties in predicting future payoffs on stocks. This makes it also very difficult to predict future market values of stocks. Therefore no-arbitrage conditions have little empirical content for the valuation of stocks, whereas they have much content for the valuation of bonds.

Being more precise, we apply the Farkas lemma to the pricing both of stock and of bonds. Let

$\tilde{R}_{it}$ = the random return of some security i in period t ($i \in I$),
I ≡ the set of all securities including stocks and bonds,
R_{ft} ≡ the risk-free rate in period t,
R_{Mt} ≡ the return of the market portfolio in period t (the market portfolio includes all risky securities issued by firms),
$\tilde{\pi}_t$ = a nonnegative valuation operator in period t that may be interpreted as a functional for pricing state-contingent claims with state-probabilities being eliminated.

Assume homogeneous expectations and a perfect capital market. Then it follows from the Farkas lemma that the capital market is arbitrage free if and only if $\tilde{\pi}_t$ exists such that

$$E(\tilde{R}_{it}) - R_{ft} = [E(\tilde{R}_{Mt}) - R_{ft}]\,\frac{\operatorname{cov}(\tilde{R}_{it}, \tilde{\pi}_t)}{\operatorname{cov}(\tilde{R}_{Mt}, \tilde{\pi}_t)}, \quad \forall\, i \in I.$$

This equation shows that the same valuation mechanism applies to stocks and bonds of all maturities. Still, this equation is of very limited use for the valuation of stocks since stock returns are governed by many different factors that appear to be unstable over time. Similarly, the market return and the valuation operator are governed by many factors so that both covariances on the right hand of the equation are hard to predict. Given the difficulties in predicting the market return and the valuation operator, the above equation has also only very limited use for the valuation of bonds.[1]

These difficulties have had very different implications for the valuation of stocks and of bonds. Since the no-arbitrage condition has little empirical content for the valuation of stocks, we usually resort to equilibrium models. The simplest model is the capital asset pricing model that can be obtained from the above equation by assuming either that $\tilde{\pi}_t$ is linear in $\tilde{R}_{Mt}$ or that all asset returns are multivariate normally distributed. Unfortunately, this model explains little if we look at empirical stock market returns. Thus, neither no-arbitrage nor equilibrium models tell us much about stock valuation.

This lack of knowledge leaves much room for assumptions on the development of stock prices over time. Thus, the standard Brownian motion could become a predominant stochastic process in stock analysis, despite research that indicates that in equilibrium Brownian motion requires very strong assumptions (see Franke [1984]; Stapleton and Subrahmanyam [1990]). Empirical evidence also indicates that stock returns do not follow a Brownian motion (see Bollerslev, Chou, and Kroner [1992], Schwert [1989]).

The route that has been followed by the valuation of bonds is quite different. Instead of considering default-free coupon bonds of different maturities, default-free zero bonds of different maturities are considered. All bonds should have a nonnegative yield to maturity in a no arbitrage world where storage of money is costless. This should be true at every date and in every state. Hence random returns of bonds in future periods must be constrained so that yields to maturity remain always nonnegative. This is a fairly strong constraint to be imposed on bond returns.

An equivalent but more easily understandable procedure is to consider the forward rates implied by the yields to different maturities. Logarithmic gross forward rates (= 1 + forward rates) are derived from differences in logarithmic gross yields to different maturities. The forward rates should also be nonnegative at every date and in each state of nature. Thus, no arbitrage imposes fairly strong constraints also on forward rates.

One implication is that bond prices cannot follow a Brownian motion. This is true since the zero bond price can never be higher than its face value. Moreover, close to maturity, the bond price must be close to its face value, and the volatility of the bond return must be very small.

The strength of the no arbitrage-constraint has reduced the need for equilibrium models for bond pricing. Yet when it comes to the question what moves bond prices over time, then we would like to know the factors driving bond prices. Those factors can be determined empirically, or they can be determined by some equilibrium model. The best of the worlds would be one in which the empirically determined factors are the same as in the equilibrium model, and the empirically observed term structure of the volatility of those factors and also the observed pricing of factor risks are consistent with that of the equilibrium model. So far, such an equilibrium model has not been developed.

One might be tempted to ignore equilibrium models given the power of the no-arbitrage models. This is dangerous, however. Since bond prices cannot follow a Brownian motion, a large variety of stochastic processes is available for choice. A purely empirical choice comes close to no-arbitrage-constrained data mining. Moreover, such a choice may be strongly time dependent and thus may have little predictive power. Therefore, there is a need for an equilibrium foundation of models for the dynamics of the term structure of interest rates.

2. Potential deficiencies of term structure models

In the following, we discuss some potential deficiencies of existing term structure models.

The problems of merging no-arbitrage and equilibrium models are easily seen when it comes to the implementation of equilibrium models. So far these models have a rather simple structure so that the implied term structure of interest rates usually is not consistent with the observed one. Using the equilibrium term structure instead of the observed one would lead to severe mispricing of bonds and interest rate derivatives. Therefore Hull and White ([1990], [1993]) propose to extend equilibrium term structures by adding some flexibility in the forward-rate process and in the process of the market price of risk. This allows them to fit the observed term structure. But this approach cannot solve the problems of inconsistency between the observed and the equilibrium term structure. The extended model of Hull and White is a no-arbitrage model and not an extended equilibrium model.

How likely are we to find a robust equilibrium model that fits the observed term structures of interest rates and of interest rate volatility? The Cox-Ingersoll-Ross model [1985] is based on rather strong assumptions, as is the Longstaff-Schwartz model [1992]. More general models are needed. To draw an analogy, consider models of exchange rate determination that presumably share some factors with term structure models. Exchange rate models so far have had very limited success. One reason for this may be that these models seem to change over time. In the early 1980s changes in monetary aggregates were considered important signals. In the mid-1980s attention shifted to current account and fiscal deficits. Today indicators of economic activity and inflation rates seem to play a major role. The nonstationarity of equilibrium models makes one rather skeptical about finding robust equilibrium models for the term structure of interest rates.

An example of a factor that appears to be important for the term structure of interest rates but has not yet been built into equilibrium models is the liquidity that varies across segments of the bond market and affects yields to maturity. Liquidity could be modeled as an option to trade such that the value of this option increases with liquidity. Such options have value, however, only if agents in the equilibrium model are to some extent heterogeneous or subject to market imperfections so that a representative investor does not exist.

Subrahmanyam [1996] has mentioned the conflict between analytical tractability of a model and consistency with stylized facts. Some examples are given below.

- Two-factor models are already quite cumbersome to handle. Whereas one-factor models may be sufficient for the analysis of short-term bonds and derivatives of them, two-factor models are the least of what is required to value options on interest-rate swaps.
- Usually normal distributions of interest rates are rejected because of the possibility of negative interest rates. Lognormal distributions avoid this problem. It is, however, questionable whether we should assume lognormality. It is more complicated to handle; moreover, portfolios of assets with lognormally distributed returns are not lognormal.

- In various models, positivity of interest rates is assured by a mean-reverting process and by a volatility that approaches zero when the interest rate approaches zero. The latter may be analytically convenient, but empirically it is doubtful. Situations in which the interest rate approaches zero may be situations in which a country faces a strong capital inflow from outside due to intensive currency speculation. Those situations are usually characterized by high volatility of short-term rates.
- Often the short-term rate is modeled by a mean-reverting process such that the short-term rate reverts to the long-term rate; in addition, a noise term is included. Now suppose that an inverse term structure of interest rates is observed. Then at any future date the probability that the term structure is still inverse is greater than 50 percent. This is in contrast to empirical evidence.
- Should term structure movements be modeled by Markov or non-Markov models? In other words, should they be modeled as path independent or as path dependent? Evidence suggests that term structure movements are heavily influenced by the business cycle. For example, if we observe a flat term structure in a strongly growing economy, then there is a rather high probability for an inverse term structure to come. If, however, we observe a flat term structure after a peak in the business cycle, then it is likely that the term structure normalizes. Hence a model that captures these effects has to be path dependent, or, alternatively, an additional state variable denoting the phase of the business cycle has to be included.

3. Implications for financial risk management

The development in the art of modeling term structure movements has important implications for financial risk management:

- A financial intermediary may use different models for different purposes. Reasons may include analytical tractability, ease of understanding, and ease of monitoring transactions. The problem with this approach is that some assets may be correctly priced using one model and over- or underpriced using another model. Hence an illusion of arbitrage opportunities is created that may easily lead to confusion.
- Since modeling interest rate risk is itself subject to various problems, the quality of risk management is reduced. Therefore, the position of the intermediary needs to be monitored and adjusted more frequently. This provides an incentive to short-termism. This short-termism can be seen easily. Bank supervisors and banks themselves are mainly concerned with the very short-term value at risk concept ignoring medium-term implications of their policy.

  Short-termism induces not only higher costs but also frequent adjustments of the bank's position purely dictated by noise in observed prices. Less frequent adjustments would avoid part of these noise-induced adjustments because from a longer-term perspective the latter are not only redundant but often have to be neutralized by subsequent reverse adjustments.
- Uncertainty about the appropriate modeling of term structure movements reduces the willingness of financial intermediaries to take interest-rate risks. This also could trigger stronger reactions of financial intermediaries to crashes and thus reinforce crashes. In any case, reduced willingness to take risks has a feedback effect on observed return processes.

4. Conclusion

Over the last twenty years, there has been impressive progress in our understanding of term structure movements and in our modeling capacities. There is no question that we shall see further progress. But at the same time, there exist various limits to our understanding of equilibrium processes. Therefore we will always face a risk of mismodeling term structure movements. This and the necessities dictated by analytical tractability constrain the reliability of risk management models.

Notes

1. The important results of Heath, Jarrow, and Morton [1992] that the drift terms in forward rates of different maturities are explained in a no-arbitrage market by their diffusion terms and by the market price of risk can be derived from the Farkas lemma. The left side of the equation above can be interpreted as the drift term, $[E(\tilde{R}_{Mt} - R_{ft}]/\text{cov}(\tilde{R}_{Mt}, \tilde{\pi}_t)$ as the market price of risk and $\text{cov}(\tilde{R}_{it}, \tilde{\pi}_t)$ as the diffusion term.

References

BOLLERSLEV, T., CHOU, R., and KRONER, K. [1992]: "ARCH Modeling in Finance: A Review of the Theory and Empirical Evidence," *Journal of Econometrics*, 52, 5–59.

COX, J.C., INGERSOLL, J.E., and ROSS, S.E. [1985]: "A Theory of the Term Structure of Interest Rates," *Econometrica*, 53, 385–467.

FRANKE, G. [1984]: "Conditions for Myopic Valuation and Serial Independence of the Market Excess Return," *Journal of Finance*, 39, 425–442.

HEATH, D., JARROW, R., and MORTON, A. [1992]: "Bond Pricing and the Term Structure of Interest Rates: A New Methodology for Contingent Claims Valuation," *Econometrica*, 60, 77–105.

HULL, J. and WHITE, A. [1990]: "Pricing Interest Rate-Derivative Securities," *Review of Financial Studies*, 3, 573–592.

HULL, J. and WHITE, A. [1993]: "One-Factor Interest Rate Models and the Valuation of Interest Rate-Derivative Securities," *Journal of Financial and Quantitative Analysis*, 28, 235–254.

LONGSTAFF, F.A. and SCHWARTZ, E.S. [1992]: "Interest-Rate Volatility and the Term Structure: A Two-Factor General Equilibrium Model," *Journal of Finance*, 47, 1259–1982.

SCHWERT, G.W. [1989]: "Why Does Stock Market Volatility Change Over Time?," *Journal of Finance*, 44, 1115–1153.

STAPLETON, R.S. and SUBRAHMANYAM, M.G. [1990]: "Risk Aversion and the Intertemporal Behavior of Asset Prices," *Review of Financial Studies*, 3, 677–693.

SUBRAHMANYAM, M.G. [1996]: "The Term Structure of Interest Rates: Alternative Approaches and Their Implications for the Valuation of Contingent Claims," *The Geneva Papers on Risk and Insurance Theory*, 21, 7–28.

The Geneva Papers on Risk and Insurance Theory, 21: 35–63 (1996)

Exotic Unit-Linked Life Insurance Contracts

STEINAR EKERN AND SVEIN-ARNE PERSSON
Norwegian School of Economics and Business Administration, Institute of Finance and Management Science, Helleveien 30, N-5035 Bergen-Sandviken, Norway

Abstract

This article integrates aspects of traditional insurance with advances in financial economics, yielding proper valuation and premium assessments of insurance benefits linked to various financial assets. Several new types of unit-linked life insurance contracts are discussed, with substantial potential for real-life applications. Compared to usual unit-linked products, these contracts offer added flexibility and/or altered exposure to financial risk for the insured and/or the insurer. The single premiums of these policies are calculated as expectations under a risk-adjusted probability measure (equivalent martingale measure), satisfying no-arbitrage conditions in financial markets.

Key words: unit-linked life insurance, exotic contracts, equivalent martingale measures, financial risk, insurance premiums

1. Introduction

This article integrates aspects of traditional insurance with recent developments in financial economics. Most traditional actuarial products have deterministic benefits, whereas some contracts in addition include profits or bonuses allocated to the policy from the surplus earned by the insurance company. A rather recent type of life insurance products links the payoff of the insurance benefits to various financial assets.

The latter type of products is the topic of this article, and we refer to these contracts as unit-linked products, though other names are used in other treatments. We give examples of new life insurance contracts based on the underlying principles of unit-linked products. The contracts are constructed by exploiting the ideas behind unit-linked products, so that either the insurer or the insured will find them more appealing. Depending on the design of the contract, one or both parties will be exposed to financial risk. The valuation of the financial risk embedded in the insurance contract will be consistent with no-arbitrage conditions in the financial markets.

In this article we treat a number of possible unit-linked contracts. We restrict attention to contracts on a single life, but the approach is really independent of how complex the insurance contract is and can easily be extended to more general insurance policies. We consider pure endowment contracts, where the benefit becomes payable if the insured survives a certain period, and term insurances, with benefit payment contingent on the insured's death within a certain period. A common life insurance contract in practice is the endowment insurance, which is a combination of a pure endowment policy and a term insurance policy.

In traditional (non-unit-linked) insurance, a deterministic benefit pattern D_t (possibly a time-invariant constant) is selected at time zero when the insurance contract is written.

For the standard, "pure vanilla" unit-linked contract, at contract initiation the benefit is irreversibly linked to one unit of a particular single mutual fund i with market value process S_t^i. More complex, "exotic" unit-linked contracts may be developed in several ways.

These types of insurances may be issued with or without a guarantee, but, due to legislation, guarantees are often included. One argument in favor of a guarantee is to protect the insured's savings. At the same time a restrictive guarantee may discourage companies from offering such products. On the other hand, to limit the insurance company's financial risk caused by exposure to large increases in the prices of the financial assets, the contracts may include a maximum value of the benefit, usually referred to as a cap.

Another extension is linking the benefit to multiple mutual funds. In all the examples the benefits are linked to one unit of one or two mutual funds, if any. Let S_t^1 and S_t^2 denote the market value at time t of mutual funds 1 and 2. If included in the insurance contract, let the guaranteed amount G_t be the minimum benefit at time t, whereas the cap K_t gives the maximum benefit at time t. The benefit payable at time t will be some function $C_t(S_t^1, S_t^2, G_t, K_t)$, where some of the arguments may be missing.

We also show several examples of flexible contracts where the unit-linked provision, if any, is selected at a fixed, future point in time τ, based on information then available, rather than when the contract is concluded. Similarly to the "chooser" or "as you like it" exotic financial option, such contracts may be referred to as "unit-linked contracts for the undecided." Just for illustrative purposes, the discussion is phrased as if the initial contract specifies a standard, simple unit-link to the first mutual fund—that is, S_t^1 is initially designated as the default benefit from time τ onward.

A possible change to a deterministic benefit G_t based on value maximization of the premium reserve will be referred to as a conversion option, presumably determined by the insured. Another possible change to a deterministic benefit K_t based on value minimization will be referred to as a buy-back option, presumably determined by the insurer. A possible unit by unit replacement of the link from the first mutual fund's market value S_t^1 to the second fund's market value S_t^2 may be termed a switch-link option. A more complex so-called choice function for future benefit can be interpreted as combinations of a switch-link option and/or a conversion option and/or a buy-back option, in which case $K_t > G_t$. Exercise of these options depend on two or more of the values $(S_\tau^1, S_\tau^2, G_\tau, K_\tau)$, which are known at the choice time τ, but where funds' market values are stochastic at the time when the contract is entered into.

The valuation of such contracts represents a challenge to the insurance industry. It appears that alternative valuation techniques are called for. Our approach is to decompose the benefit pattern into simple components of increasing complexity, which can be valued separately. By the market value additivity principle, the value of the whole is equal to the sum of the values of the parts. Fortunately, several of the benefit structures are somewhat similar to so-called exotic options in the finance literature. If so, we may call upon established valuation results in the option pricing literature (Black and Scholes [1973], Margrabe [1978], Stulz [1982]) as elements in the valuation of complex exotic unit-linked insurance contracts.

A main contribution of the article is the illustration of how to compute appropriate single premiums as market values of the exotic unit-linked contracts, combining financial risk and mortality risk. Actuarial risk is treated as in the classical actuarial models, whereas financial risk aversion is incorporated consistently with market price processes of financial assets. In our perfect economy context, we also assume away any transaction costs, administrative costs, or other frictions.

The initial valuation of such exotic unit-linked insurance contracts is seen to depend on a number of factors. The relevant financial parameters are the time-zero market values of the funds, parameters related to the funds' variances and covariances, and the interest rate. The contract design specifies the guarantee profile, the cap profile, and the relevant points in time. These two sets of factors determine the economic valuation of the benefits, the overall financial risk exposure of the contract, and how the financial risk is distributed between the two involved parties. Finally, the total value of the insurance contracts is determined by including mortality factors.

The first treatments of unit-linked contracts with guarantees of which we are aware, are Brennan and Schwartz [1976, 1979a, 1979b] and Boyle and Schwartz [1977]. They recognized that the payoff of a unit-link insurance with a guarantee is related to the payoffs of certain financial options and applied the option pricing theory initiated by the results of Black and Scholes [1973].

Later papers (Delbaen [1990], Bacinello and Ortu [1993a, 1993b, 1994, 1995], Nielsen [1993], Nielsen and Sandmann [1995], Aase and Persson [1994] and Persson [1994]) applied the martingale-pricing theory credited to Harrison and Kreps [1979]. The main achievement by this methodology is that the single premium of an insurance contract may be calculated as an expectation under a risk-adjusted probability measure.

The method of calculating single premiums as expectations has its own name in the actuarial literature, the principle of equivalence, but observe carefully that this classical principle does not include any adjustment for risk aversion. The classical actuarial models do not deal with financial risk, and the approach of changing the probability measure to obtain a proper adjustment for financial risk is a rather recent one in the actuarial literature.

Whereas the focus of most earlier works is the valuation of a guarantee connected to a unit-linked contract, we also look at unit-linked contracts with other components such as caps and certain switch dates and show connections between these contracts and contracts with or without guarantees. We apply the Harrison and Kreps [1979] methodology with a fixed interest rate and a continuous model of mortality. Periodical premiums and problems related to the actual investment behavior in the financial markets are not discussed.

In addition to pricing methodology, the literature on unit-linked insurance can be categorized according to model of mortality, treatment of interest rates, and periodical premium/investment behavior.

Aase and Persson [1994] and Persson [1994] apply a continuous-time model of mortality. Other studies discretize mortality on an annual basis. As a consequence, the two works just mentioned establish a connection between the classical Thiele equation of the actuarial sciences and the Black and Scholes [1973] equation, which was the basis for the option-pricing-related results.

An important topic of interest for the insurance industry is the future development of the interest rate. Bacinello and Ortu [1993b, 1994, 1995], Nielsen [1993], Nielsen and Sandmann [1995], and Persson [1994] include a stochastic model of the interest rate when dealing with unit-linked contracts. The other works assume a deterministic development of the interest rate.

A number of problems connected to the periodical premiums and the corresponding investment behavior in the financial markets are discussed in Brennan and Schwartz [1976, 1979a, 1979b] and Boyle and Schwartz [1977], Delbaen [1990], Bacinello and Ortu [1993a, 1993b, 1994, 1995], Nielsen [1993], Nielsen and Sandmann [1995], and Aase and Persson [1994].

To make this article self-contained and more accessible to insurance economists who are not necessarily quite familiar with the modern finance literature, we include a description of a model of a financial market and explain how we use economic valuation theory in pricing the building blocks.

The article is structured as follows: Section 2 presents the economic model, stressing valuation of insurance contracts according to the martingale approach. Section 3 illustrates how various one-fund unit-linked benefit structures yield different market values, from which single premiums of the complete insurance policies can be calculated. Section 4 extends the analysis to benefit patterns depending on multiple funds—for simplicity two funds. Section 5 considers further extensions, where the link characteristics are determined at a fixed future date and are thus random when the insurance contracts are entered into. Section 6 contains some concluding remarks. Proofs of some lemmas can be found in an appendix.

2. Valuation and economic model

2.1. *Valuation*

The insurance benefit is linked to a set S^I of risky financial assets, usually a set of mutual funds. The market value at time t of one unit of mutual fund i is denoted by S_t^i. Let $S_t^I :=$ $\{S_t^i : t \geq 0\}$ (where := signifies defined to be equal) be the joint stochastic process for the market values of the mutual funds. In case of just one mutual fund, the superscripts I and i are usually omitted. The insurance benefit may also depend on a set D^J of deterministic benefit patterns from traditional insurance. The insurance benefit payable at time t is some function $C_t = C_t(S_t^I, D_t^J)$, but some or all of the arguments are often suppressed. Observe that the benefit functional form itself, and not just the argument values, may depend on the time t. The mutual funds do not pay any dividends, so all returns occur as capital gains.

The process $\{r_t\colon t \geq 0\}$ represents the short-term interest rate. The quantity

$$v(t) := \exp\left(-\int_0^t r_s ds\right) \tag{1}$$

is sometimes called the discount function and represents the present value at time zero of one currency unit at time t.

All random variables are defined on a given probability space $(\Omega, \mathcal{F}, P)$. Here r_t and S_t are measurable with respect to a filtration $\{\mathcal{G}_t\colon t \geq 0\}$. Let the random variable T_x represent the remaining lifetime of the insured (customarily assumed to depend only on x, the insured's age at the point of issue), and let T_x be measurable with respect to a sigma algebra $\mathcal{H}$. The sigma algebras $\mathcal{H}$ and $\mathcal{G}_t$ are assumed to be independent, reflecting that the remaining life time of the insured is independent of the interest rate and the financial securities. The stochastic processes $\{S_t^i: t \geq 0\}$ and $\{r_t: t \geq 0\}$ and the random variable T_x generate all uncertainty in the model, so the total information available at time t is given by a filtration $\{\mathfrak{F}_t: t \geq 0\}$, where $\mathfrak{F}_t$ represents the sigma algebra union of $\mathcal{G}_t$ and $\mathcal{H}$. We impose a fixed time horizon T, so that the sigma algebra $\mathfrak{F} = \mathfrak{F}_T$. Finally, $\mathfrak{F}_0 = \{\emptyset, \Omega\}$, meaning that no relevant uncertainty has been resolved at time zero. In particular, the conditional expectation with respect to $\mathfrak{F}_0$ is equivalent to the unconditional expectation.

We present unit-linked formulas both for pure endowment contracts and for term insurance contracts. Our results for pure endowment policies can accordingly be combined with the results for term insurances as well as traditional term insurances to create interesting endowment policies.

The expression for the single premium π of an insurance policy issued at time zero will be of the following form

$$\pi = E^Q[v(U)C_U]. \tag{2}$$

The random variable U represents the random time to expiration (the typical example will be the remaining life time of the insured). The random amount $C_U = C_U(S_U)$ represents the insurance benefit, which is a function of the market value of a mutual fund at the time that the benefit expires. The product inside the expectation operator is thus the present value of the random amount that the insured receives at the future uncertain time U.

The single premium π is simply the expected present value of this random insurance benefit. Note carefully, however, that this expectation is calculated with respect to a risk-adjusted probability measure Q constructed from economic theory.

In general, the transformation from the actual probability measure P, governing the occurrence of future events, to the constructed measure Q represents the correct, in a certain sense, adjustment for both financial risk and mortality risk.

Whereas mortality risk is an important issue in potential applications of this theory, we see few reasons to treat mortality risk connected to the unit-linked contracts different from mortality risk connected to the traditional life insurance contracts. With respect to pricing, mortality risk is therefore treated as unsystematic risk, meaning that the insurer does not receive any economic compensation for accepting that risk. This approach is also implicit in the traditional principle of equivalence. It is justified by the traditional argument saying that the insurer can, at least in principle, eliminate mortality risk by increasing the number of identical policies in his portfolio.

Hence, there are no needs for risk adjustment of mortality risk. Thus, with some slight abuse of notation, in the remainder of the article the transformation from the measure P to Q represents adjustment for financial risk only. We refer to Aase and Persson [1994] and Persson [1994] for a discussion.

By conditioning with respect to T_x, which is customarily assumed to be a continuous, nonnegative random variable, expression (2) may be written

$$\pi = \int_0^\infty E^Q[v(t)C_t]f_x(t)dt, \tag{3}$$

where $f_x(t)$ represents the (known) probability density of T_x. This expression depends crucially on the assumed independence between the financial factors and the insured's remaining lifetime. As indicated above, the probability measure Q in equation (3) is based on risk adjustment of financial risk only, whereas the Q of equation (2) in principle incorporates adjustments for both mortality risk and financial risk.

2.2. The Black and Scholes economy

We now specialize the processes for the financial variables. First we assume that the short interest rate is constant—that is, $r_t = r$ for all t. We notice that the discount function

$$v(t) = e^{-rt}$$

is deterministic. We introduce a Brownian motion, $\{W_t: t \geq 0\}$ on $(\Omega, \mathfrak{F}, P)$. The price process for the risky fund (the extension to more than one such fund follows in the next subsection) is assumed to be a geometric Brownian motion,

$$dS_t = \eta S_t dt + \sigma S_t dW_t, \tag{4}$$

where S_0, η, and σ are given constants interpretable as the initial value of the fund, the instantaneous expected rate of return of S_t, and the instantaneous standard deviation of the rate of return of S_t, respectively. We will some places refer to η as the drift parameter and σ as the volatility parameter. In applications one may estimate η and σ, as well as r, from market data.

Let Q be the probability measure defined by

$$\frac{dQ}{dP} = \exp\left(-\left(\frac{\eta - r}{\sigma}\right) W_T - \frac{1}{2}\left(\frac{\eta - r}{\sigma}\right)^2 T\right). \tag{5}$$

It is easy to verify that this quantity is a nonnegative random variable with unit expectation and finite variance, thus the measure Q is well defined. It follows from Girsanov's theorem that the price process for S_t under the equivalent probability measure Q is

$$dS_t = rS_t dt + \sigma S_t d\hat{W}_t,$$

where $\hat{W}_t$ is a Brownian motion on $(\Omega, \mathfrak{F}, Q)$. The solution to this stochastic differential equation is

$$S_t = S_0 e^{(r-(1/2)\sigma^2)t + \sigma\hat{W}_t} \tag{6}$$

an expression that does not depend on η, the drift parameter of the mutual fund.

Recalling that the discount function $v(t) = e^{-rt}$, the process $\{v(t)S_t: t \geq 0\}$ is called a discounted price process. From standard properties of normally distributed random variables it follows that $E^Q[v(s)S_s \mid \mathfrak{F}_t] = v(t)S_t$, for $s \geq t$, taking expectations under the probability measure Q.

Lemma 2.1. *The discounted price process* $\{v(t)S_t: t \geq 0\}$ *is a martingale with respect to the probability measure* Q *and the filtration* $\{\mathfrak{F}_t: t \geq 0\}$. □

In particular, observe that $E^Q[v(t)S_t] = S_0$ for $t \in [0, T]$.

2.3. *Extended financial model*

Some of the policies we consider involve more than one risky asset. In the current treatment the number of risky assets is limited to two. The model is therefore extended accordingly. The price processes of the two mutual funds are given by the following stochastic differential equations,

$$dS_t^1 = \eta_1 S_t^1 dt + \sigma_1 S_t^1 dW_t^1 + \sigma_2 S_t^1 dW_t^2 \tag{7}$$

and

$$dS_t^2 = \eta_2 S_t^2 dt + \sigma_3 S_t^2 dW_t^1 + \sigma_4 S_t^2 dW_t^2,$$

where the constants η_1 and η_2 have similar interpretations as in the one-dimensional case, and the initial values S_0^1 and S_0^2 are given. We may interpret $\sqrt{\sigma_1^2 + \sigma_2^2}$ as the instantaneous standard deviation of the return of S_t^1, $\sqrt{\sigma_3^2 + \sigma_4^2}$ as the instantaneous standard deviation of the return of S_t^2 and $(\sigma_1\sigma_3 + \sigma_2\sigma_4)$ as the instantaneous covariance between the rate of return of S_t^1 and S_t^2. Here W_t^1 and W_t^2 are two independent Brownian motions defined on $(\Omega, \mathfrak{F}, P)$. Now define

$$\lambda_1 = \frac{\eta_2\sigma_2 - \eta_1\sigma_4 - r(\sigma_2 - \sigma_4)}{\sigma_3\sigma_2 - \sigma_1\sigma_4}$$

and

$$\lambda_2 = \frac{\eta_1\sigma_3 - \eta_2\sigma_1 - r(\sigma_3 - \sigma_1)}{\sigma_3\sigma_2 - \sigma_1\sigma_4}.$$

Both λ_1 and λ_2 are constants. We now define a probability measure Q by

$$\frac{dQ}{dP} = \exp\left(-\lambda_1 W_T^1 - \lambda_2 W_T^2 - \frac{1}{2}(\lambda_1^2 + \lambda_2^2)T\right). \tag{8}$$

Also here dQ/dP is a strictly positive random variable with unit expectation and finite variance. The price processes under Q are now

$$dS_t^1 = rS_t^1 dt + \sigma_1 S_t^1 d\hat{W}_t^1 + \sigma_2 S_t^1 d\hat{W}_t^2$$

and

$$dS_t^2 = rS_t^2 dt + \sigma_3 S_t^2 d\hat{W}_t^1 + \sigma_4 S_t^2 d\hat{W}_t^2$$

where the "hats" indicate that the Brownian motions are defined on $(\Omega, \mathfrak{F}, Q)$. By solving this system under Q we obtain

$$S_t^1 = S_0^1 \exp\left(\left(r - \frac{1}{2} V_x^2\right) t + X_t\right) \tag{9}$$

and

$$S_t^2 = S_0^2 \exp\left(\left(r - \frac{1}{2} V_y^2\right) t + Y_t\right),$$

where

$$X_t := \sigma_1 \hat{W}_t^1 + \sigma_2 \hat{W}_t^2$$

and

$$Y_t := \sigma_3 \hat{W}_t^1 + \sigma_4 \hat{W}_t^2.$$

Furthermore, X_t and Y_t are binormally distributed with

$$V_x^2 := \frac{1}{t} \operatorname{Var}(X_t) = \sigma_1^2 + \sigma_2^2,$$

$$V_y^2 := \frac{1}{t} \operatorname{Var}(Y_t) = \sigma_3^2 + \sigma_4^2,$$

and

$$\Gamma := \frac{1}{t} \operatorname{Cov}(X_t, Y_t) = \sigma_1 \sigma_3 + \sigma_2 \sigma_4.$$

We define

$$\Psi^2 = V_y^2 + V_x^2 - 2\Gamma. \tag{11}$$

From the expressions labeled (10) we see that Ψ^2 represents $1/t$ times the variance of the difference between the random variables X_t and Y_t, which again is closely related to the difference between the market values of the two assets.

Later we will present our results in terms of the quantities V_x, V_y, Γ, and Ψ.

The following lemma extends Lemma 2.1.

Lemma 2.2. *The discounted price processes* $\{v(t)S_t^1\colon t \geq 0\}$ *and* $\{S_t^2 v(t)\colon t \geq 0\}$ *are martingales with respect to the probability measure* Q *and the filtration* $\{\mathfrak{F}_t\colon t \geq 0\}$. □

2.4. Arbitrage pricing results

An equivalent martingale is a probability measure Q that is equivalent to P, such that the discounted price processes are martingales with respect to this measure and such that the Radon-Nikodym derivative has finite variance.

An arbitrage opportunity is a trading strategy in the financial market which generates a positive (nonnegative) cash flow from a nonpositive (negative) initial investment. The absence of arbitrage opportunities is a mild requirement for a meaningful economic model.

The following powerful result is central in modern financial economics and is part of the so-called fundamental theorem of asset pricing.

Lemma 2.3. *The existence of an equivalent martingale measure is equivalent to no arbitrage opportunities in the economy.* □

This lemma shows the connection between the martingale property of the discounted asset prices, which is a purely mathematical property, and the absence of arbitrage opportunities, which is a meaningful economic condition.

Furthermore, the two described economies are complete in the sense that any random payoff with finite variance has a unique market price.

From Lemmas 2.1 and 2.2, the discounted price processes are martingales. Lemma 2.3 then establishes that the two described economies are arbitrage free, which serves as a justification for the use of the martingale-based arbitrage pricing theory. This theory is by now standard in financial economics. We refer the interested reader to the seminal paper of Harrison and Kreps [1979], advanced textbooks in finance such as Duffie [1992], and the survey article by Dybvig and Ross [1992].

2.5. Market-based single premiums

Based on the economic model described above, pricing formulas for various unit-linked insurance contracts may be derived. Mainly to keep the notation simple, we present the market prices of the policies at time zero.

It is convenient to introduce the following quantity

$$V_t := E^Q[v(t)C_t] = E^Q[e^{-rt}C_t] \tag{12}$$

to be interpreted as the market value at time zero of a random benefit that is paid at time t with probability one (superscripts will be used to distinguish different benefits). Evaluation of this term is an important building block for the market values of the insurance contracts treated here, as demonstrated by Propositions 2.1 and 2.2 below.

Recall the key assumption of independence between mortality and the financial market. Substitution of expression (12) into expression (3) yields the single premium

$$\pi = \int_0^\infty V_t f_x(t)\,dt. \tag{13}$$

Let T denote the term of the contract and ${}_tp_x$ and the survival probability—that is,

$$ {}_tp_x := P(T_x > t) = \int_t^\infty f_x(s)\,ds. $$

The single premiums, of the various pure endowment and term insurances, will be of the forms presented in the following proposition.

Proposition 2.1. *The single premiums at time zero, π and π^1, of the pure endowment insurance and the term insurance, expressed by the quantity V_t, are given by*

$$ \pi = {}_Tp_x V_T \tag{14} $$

and

$$ \pi^1 = \int_0^T V_t f_x(t)\,dt, \tag{15} $$

respectively. □

Proof: The results follow from equation (13). For the term insurance, the upper limit of integration is the fixed term of the insurance T. For the pure endowment contract with fixed expiry at T, $V_t = 0$ for $t < T$ and $V_t = V_T$ otherwise. □

Whereas superscripts will be used on V_t to distinguish different benefits, following traditional actuarial notation, superscript 1 on π indicates that the benefit is payable on death. Instead subscripts on π will be used, when needed, to distinguish the various contracts. In some cases π^* will be written for the single premium of an endowment police, having both pure-endowment and term-insurance components.

2.6. *Premiums of flexible contracts*

With the most general benefit functions $C_t = C_t(S_t^I, D_t^J)$, the benefit may depend on the current market values of a number of risky assets and on a number of deterministic schedules, and with functional forms possibly shifting over time. With standard, elementary unit-linked contracts, the benefit $C_t = C_t(S_t^i)$ depends on one particular risky asset i. With traditional, elementary contracts, the benefit $C_t = D_i^{\,j}$ depends on one particular deterministic pattern j. In both latter cases, the choice of benefit linkage to one single process is irreversibly made at time zero.

Flexible unit-linked contracts are somewhere between the two polar cases, having the feature that the ultimate link characteristics of the contracts are determined at some fixed, future point in time τ. The link choice at time τ follows from some choice function agreed on at time zero, as well as information first available at time τ. At that point in time, the insurance contract will be irrevocably linked to one particular, univariate process. The selected link basis will be either some risky asset with price process $S_t^i \in S_t^I$ or some deterministic asset value process $D_t^j \in D_t^J$. Let $\Theta_\tau(\cdot \mid \mathfrak{F}_\tau)$ denote the choice function, conditional on the information set $\mathfrak{F}_\tau$, and yielding one single process. Thus, for the time interval $[\tau, T]$, the insurance benefit is $C_t = C_t(\Theta_\tau(S_t^I, D_t^J \mid \mathfrak{F}_\tau))$. The exact nature of the choice function will be left unspecified until Section 5.

Let

$$V_{\tau,t} := E^Q[e^{-(t-\tau)r} C_t \mid \mathfrak{F}_\tau]$$

be the market value at time τ of a certain benefit C_t payable at time $t \geq \tau$. By conditioning on the sigma algebra $\mathfrak{F}_\tau$, expression (12) for the time-zero market value yields

$$V_t = E^Q[e^{-r\tau} V_{\tau,t}], \text{ for } t \geq \tau.$$

Observe that the discounting interval has changed from $[0, t]$ to $[0, \tau]$, and that the future benefit C_t has been replaced by its market value $V_{\tau,t}$ at time τ, which is random at time zero. The general expression (13) for the single premium can thus be written as

$$\pi = \int_0^\tau V_t f_x(t)\,dt + \int_\tau^T E^Q[e^{-r\tau} V_{\tau,t}] f_x(t)\,dt. \tag{16}$$

Next suppose the discounted benefit process $\{e^{-(t-\tau)r} C_t : t \geq \tau\}$ is a martingale with respect to the probability measure Q and the filtration $\mathfrak{F}_t : t \geq \tau$. Then

$$V_{\tau,t} = E^Q[e^{-(t-\tau)r} C_t \mid \mathfrak{F}_\tau] = C_\tau, \text{ for all } t \geq \tau.$$

Under the martingale assumption, the time-zero market value of the benefit C_t is

$$V_\tau^* : = E^Q[e^{-r\tau} C_\tau], \text{ for all } t \geq \tau.$$

Note that we do not assume that the process before time τ—that is, $\{e^{-rt} C_t : \tau > t \geq 0\}$—is a martingale, and thus V_τ^* may be different from C_0. Putting V_τ^* outside the integral, the general expression (16) for the single premium can then be written as

$$\pi = \pi^A + V_\tau^* \int_\tau^\infty f_x(t)\,dt,$$

where

$$\pi^A := \int_0^\tau V_t f_x(t)\,dt$$

is the premium attributed to some arbitrary benefit function applicable to the initial, pre-choice time interval $[0, \tau)$. The exact nature of this initial benefit is not a central issue in our discussion and is therefore left unspecified, but Sections 3 and 4 illustrate some possibilities.

The premium reserve at time t of a policy represents the market value of the policy at time t. It can also be interpreted as the insurer's debt to the insured at time t or as the necessary funds the insurer should reserve at time t for future net claims. The premium reserves at time t for the pure endowment and the term insurance, conditioned on the insured being alive at time t, can be expressed as

$$\pi_t = {}_{T-t}p_{x+t} V_{t,T}$$

and

$$\pi_t^1 = \int_0^{T-t} V_{t,t+s} f_{x+t}(s)\, ds,$$

respectively, where $f_{x+\tau}(\cdot)$ represents the probability density of the remaining life time of an $(x + \tau)$-year-old person. From the relation $f_{x+\tau}(t - \tau) = f_x(t)/{}_tp_x$, $t \geq \tau$, one may interpret $f_{x+\tau}(\cdot)$ as the conditional probability density function of the remaining life time in τ years of a currently x-year-old person, given that he will survive the first τ years.

Proposition 2.2. *Under the martingale assumption with respect to the discounted benefit process $\{e^{-(t-\tau)r}C_t\colon t \geq \tau\}$, the single premiums at time zero, π and π^1, of the pure endowment insurance and the term insurance, expressed by the quantity V_τ^*, are given by*

$$\pi = {}_Tp_x V_\tau^* \tag{17}$$

and

$$\pi^1 = \pi^A + [{}_\tau p_x - {}_Tp_x]V_\tau^* = \pi^A + {}_\tau p_x[1 - {}_{T-\tau}p_{x+\tau}]V_\tau^*, \tag{18}$$

respectively. □

Proof: For the pure-endowment insurance expiring at T, the time-zero market value of the benefit is $V_\tau^* = E^Q[e^{-r\tau}C_\tau] = E^Q[e^{-r\tau}E^Q[e^{-(T-\tau)r}C_T|\mathfrak{F}_\tau]] = E^Q[e^{-rT}C_T] = V_T$, payable with the survival probability ${}_Tp_x$. The premium component π^A covers the cost associated with possible payment of the arbitrary benefit in case of death before the link choice at τ. The term insurance benefit selected at time τ, with time-zero present value of $V_\tau^* = E^Q[e^{-r\tau}C_\tau] = E^Q[e^{-r\tau}E^Q[e^{-r\tau}E^Q[e^{-(t-\tau)r}C_t|\mathfrak{F}_\tau]] = E^Q[e^{-rt}C_t] = V_t$, for $T \geq t \geq \tau$, will be payable with probability

$$\int_\tau^T f_x(s)\, ds = \int_\tau^\infty f_x(s)\, ds - \int_T^\infty f_x(s)\, ds,$$

which simplifies to the term $[{}_\tau p_x - {}_Tp_x]$ of the first equality of (18). The second equality follows from substitution of the survival probability relationship ${}_Tp_x = {}_\tau p_x\, {}_{t-\tau}p_{x+\tau}$ and factoring out ${}_\tau p_x$. □

For a standard unit linked contract, without any delayed link choice, $\tau = 0$. Hence, trivially ${}_{\tau}p_x = 1$ and ${}_{T}p_x = {}_{T-\tau}p_{x+\tau}$. Proposition 2.2 and anticipating Lemma 3.1 yield the time-zero premiums $\pi = {}_{T}p_x S_0$ for pure-endowment insurance and $\pi^1 = [1 - {}_{T}p_x]S_0$ for time insurance. For a traditional insurance, with deterministic benefits $D_t = D_0 e^{rt}$ escalated over time at the risk-free rate of interest, Proposition 2.2 yields the time-zero premiums $\pi = {}_{T}p_x D_0$ for pure endowment insurance and $\pi^1 = [1 - {}_{T}p_x]D_0$ for term insurance.

The results of Propositions 2.1 and 2.2 are based on the characteristics of the contracts, our economic model, and the assumed independence between financial and mortality factors. These results are therefore our basic valuation tools. In the following sections we specify a number of different benefits and calculate their corresponding market values at time zero.

3. One-fund unit-linked insurance contracts

Below a series of contracts are discussed. In each case, the benefit C_t is described. All such benefits are linked to one single mutual fund, with price S_t at time t. The contracts are of increasing complexity.

In the first four subsections, the time-zero market of the benefit $V_t = E^Q[v(t)C_t]$, is derived and presented as a lemma. Applying Proposition 2.1 to these lemmas will then give the single premiums of the contracts, either pure endowment ones or term insurance ones, with the respective benefits.

In the subsequent two subsections, the pure endowment and the term insurance components are combined into one unit-linked endowment insurance policy. The market value of the two endowment contracts with quite different unit-link characteristics are derived and compared.

3.1. Standard unit-linked insurance

With the standard, "pure vanilla" contract, the unit-linked benefit at time t is just

$$C_t^1 = S_t \tag{19}$$

—that is, one unit of the mutual fund. For completeness, we state its time zero market value S_0 as a lemma, (following immediately from Lemma 2.1):

Lemma 3.1. *The market value at time zero of the benefit* $C_t^1 = S_t$ *is*

$$V_t^1 = S_0. \tag{20}$$ □

If the mutual fund were to pay dividends, then V_t^1 would have been S_0 minus the present value of future dividends. Several of the following expressions, properly modified, would go through with dividends as well. However, for simplicity we assume no dividend payments.

3.2. Unit-linked insurance with guarantee

In this example the benefit of the policy is one unit of the mutual fund, but in order to protect the insured a guarantee is included in the contract. Thus the benefit payable at time t may be written

$$C_t^2 = \max[S_t, G_t]. \tag{21}$$

Lemma 3.2. *The market value at time zero of the benefit* $C_t^2 = \max[S_t, G_t]$ *is*

$$V_t^2 = S_0\Phi(d_1^t) + G_t e^{-rt}\Phi(-d_2^t), \tag{22}$$

where $\Phi(\cdot)$ *is the standard normal distribution function,*

$$d_1^t := \frac{1}{\sigma\sqrt{t}}\left[\ln\left(\frac{S_0}{G_t}\right) + \left(r + \frac{1}{2}\sigma^2\right)t\right] \tag{23}$$

and

$$d_2^t := \frac{1}{\sigma\sqrt{t}}\left[\ln\left(\frac{S_0}{G_t}\right) + \left(r - \frac{1}{2}\sigma^2\right)t\right] \qquad \square$$

Thus, the time-zero market value V_t^2 of the benefit can be decomposed into two terms, corresponding to the first term of the Black and Scholes European call option and put option pricing formulas, respectively.

3.3. Capped benefits

To limit the exposure to large upside price increases, the insurer may want to limit the maximum value of the benefit to an amount K_t, often referred to as a cap. In the financial literature a number of papers deal with pricing of capped options, which are specialized financial instruments that have similar payoffs to this insurance contract (see, e.g., Broadie and Detemple [1995]).

Thus, we consider the benefit payable at time t

$$C_t^3 = \min[S_t, K_t]. \tag{24}$$

Observe that the benefit can be rewritten in various ways. From $C_t^3 = S_t - \max[S_t - K_t, 0]$, the benefit equals the underlying asset minus the payoff of a call option. From $C_t^3 = K_t - \max[K_t - S_t, 0]$, the benefit corresponds to the cap minus a put option payoff. From $C_t^3 = (S_t + K_t) - \max[S_t, K_t]$, the benefit can be interpreted as the sum of the underlying fund value and the cap, minus the benefit C_t^2 of the previous subsection.

Lemma 3.3. *The market value at time zero of the benefit* $C_t^3 = \min[S_t, K_t]$ *is*

$$V_t^3 = S_0\Phi(-d_3^t) + K_t e^{-rt}\Phi(d_4^t), \tag{25}$$

where

$$d_3^t := \frac{1}{\sigma\sqrt{t}}\left[\ln\left(\frac{S_0}{K_t}\right) + \left(r + \frac{1}{2}\sigma^2\right)t\right] \tag{26}$$

and

$$d_4^t := \frac{1}{\sigma\sqrt{t}}\left[\ln\left(\frac{S_0}{K_t}\right) + \left(r - \frac{1}{2}\sigma^2\right)t\right]. \quad \square$$

3.4. Capped benefits with a guarantee

With both a cap K_t and a guaranteed amount G_t, the benefit payable at time t becomes

$$C_t^4 = \min[\max[S_t, G_t], K_t], \tag{27}$$

with $G_t < K_t$. Compared to a simple unit-linked payoff, this structure of the benefit reduces both the downside risk and the upside potential for the insured. The insurer is also exposed to a limited financial risk compared to the case of a simple unit-linked contract.

Lemma 3.4. *The market value at time zero of the benefit* $C_t^4 = \min[\max[S_t, G_t], K_t]$ *is*

$$V_t^4 = G_t e^{-rt}\Phi(-d_2^t) + S_0[\Phi(d_1^t) - \Phi(d_3^t)] + K_t e^{-rt}\Phi(d_4^t). \tag{28}\square$$

When G_t approaches zero—that is, a situation without any guarantee—V_t^4 tends to V_t^3. Similarly, when K_t approaches infinity—that is, a situation without any cap—V_t^4 tends to V_t^2.

3.5. Simple endowment insurance with deferred term insurance

An endowment insurance is a combination of a pure endowment policy and a term insurance policy. Whereas the benefit from the term insurance component is normally payable at death, we shall in the remainder of this section consider a policy where the payment of the term insurance benefit is deferred until the end of the insurance period. Thus, the benefit from either the pure endowment component or the term insurance component are payable at a common, fixed expiration date T. An example may be an insured who receives a pure endowment benefit at his retirement age of sixty-seven, if he is alive. If he is dead, his heirs receive the term insurance benefit at the time at which he would have been sixty-seven.

Suppose first the pure endowment benefit is a standard unit-linked one, with $C_T^5 = G_T$. The term insurance is a traditional one yielding a fixed amount C_T^5, except that the contingent payment is payable at time T rather than at death. The survival probability is ${}_Tp_x$.

Proposition 3.1. *Consider the endowment insurance with deferred term insurance, such that the benefit at time T is $C_T^5 = S_T$ if the insured is alive and $C_T^5 = G_T$ otherwise. The time-zero market value of this contract is*

$$\pi_5^* = {}_Tp_x S_0 + (1 - {}_Tp_x)G_T e^{-rT}. \tag{29}\square$$

For the special case ${}_Tp_x = 1/2$, this expression simplifies into $\pi_5^* = (1/2)[S_0 + G_Te^{-rT}]$ —that is, one-half of the sum of the current value of the mutual fund and the discounted value of the fixed amount.

3.6. *Exotic endowment insurance with deferred term insurance*

Next suppose the pure endowment benefit is unit-linked with a fixed guarantee G_T, such that $C_T^6 = C_T^2 = \max[S_T, G_T]$. The term insurance is both deferred and unit-linked but capped at the same amount G_T, giving $C_T^6 = C_T^3 = \min[S_T, G_T]$. Compared with the simple endowment contract of the previous subsection, the unit-linked pure endowment component is more attractive to the insured, and the term insurance component is less attractive to the insured. Both components are less risky than similar simple unit-linked contracts, without the guarantee and cap feature.

Proposition 3.2. *Consider the endowment insurance with deferred term insurance, such that the benefit at time T is* $C_T^6 = \max[S_T, G_T]$ *if the insured is alive and* $C_T^6 = \min[S_T, G_T]$ *otherwise. The time zero market value of this contract is*

$$\pi_6^* = [2{}_Tp_x - 1][S_0\Phi(d_1^T) + G_Te^{-rT}\Phi(-d_2^T)] + [1 - {}_Tp_x][S_0 + G_Te^{-rT}]. \quad (30)\square$$

Of the terms in the expression of Proposition 3.2, $[2{}_Tp_x - 1]$ and $[1 - {}_Tp_x]$ reflect mortality factors, S_0 and G_Te^{-rT} are current market values of S_T and G_T, respectively, and $S_0\Phi(d_1^T)$ and $G_Te^{-rT}\Phi(-d_2^T)$ appear in the Black and Scholes call and put option pricing formula, respectively.

For the special case of the survival probability ${}_Tp_x = 1/2$, Proposition 3.2 shows that the single premium simplifies into $\pi_6^* = (1/2)[S_0 + G_Te^{-rT}]$. Thus, for this special mortality structure, the single premiums for the two endowment policies with deferred term insurance are equal, $\pi_5^* = \pi_6^*$. The reason is that the valuation of the risk reductions caused by the guarantee and by the cap exactly cancel out.

For other survival probabilities, however, the premiums from the complex and simple endowment unit-linked policies are different.

Proposition 3.3. *The difference between the single premium* π_6^* *of an exotic endowment insurance consisting of a guaranteed unit-linked pure endowment policy and a deferred and capped term insurance, and the single premium* π_5^* *of an endowment insurance consisting of a simple unit-linked pure endowment insurance and a deferred but otherwise traditional term insurance, is given by*

$$\pi_6^* - \pi_5^* = [2{}_Tp_x - 1][G_Te^{-rT}\Phi(-d_2^T) - S_0\Phi(-d_1^T)]. \quad (31)\square$$

The proposition confirms that the two endowment policies with term insurance have the same single premiums for ${}_Tp_x = 1/2$.

As $d_1^T > d_2^T$, $\Phi(-d_2^T) > \Phi(-d_1^T)$. Thus, the rightmost expression in square brackets is positive for $G_Te^{-rT} > S_0$, which in option jargon corresponds to a European put option being "in the money" and the call option being "out of the money." If, in addition ${}_Tp_x >$ $(<)$ 1/2, then $\pi_6^* > (<)\ \pi_5^*$.

4. Multifund unit-linked insurance contracts

In the one-fund unit-linked insurance contracts of the previous section, the insurance benefit of each contract was linked to the market value $S_t = S_t^i$ of a particular mutual fund, when the contract was issued. In contrast, in this section the unit-linked insurance contracts will at time zero be linked to the joint performance of the set S^I of multiple funds, with market value process S_t^I. We restrict attention to cases where any contract will be linked to at most two different mutual funds—that is, $S_t^I = \{S_t^1, S_t^2\}$. The extension of linkage to more than two funds would considerably complicate computations. Guarantee processes G_t and cap processes K_t may be included in the multifund contracts as well.

Below we describe the benefit patterns $C(S_t^1, S_t^2, G_t, K_t)$ of various insurance contracts. The time-zero market values V_t of each benefit are presented in lemmas. Using these benefit market values in Proposition 2.1 then gives the time-zero single insurance premiums π and π^1, for the pure endowment and term insurance contracts, respectively.

4.1. Simple two-fund contracts

As an introduction we consider the contract where the insured receives one unit of the fund of the highest value—that is,

$$C_t^7 = \max[S_t^1, S_t^2]. \tag{32}$$

Lemma 4.1. *The market value at time zero of the benefit* $C_t^7 = \max[S_t^1, S_t^2]$ *is*

$$V_t^7 = S_0^1\Phi(\beta_1^t) + S_0^2\Phi(-\beta_2^t), \tag{33}$$

where

$$\beta_1^t := \frac{1}{\Psi\sqrt{t}}\left[\ln\left(\frac{S_0^1}{S_0^2}\right) + \frac{1}{2}\Psi^2 t\right] \tag{34}$$

and

$$\beta_2^t := \frac{1}{\Psi\sqrt{t}}\left[\ln\left(\frac{S_0^1}{S_0^2}\right) - \frac{1}{2}\Psi^2 t\right]. \qquad \square$$

This formula does not depend on the interest rate r, as its effect as a discounting rate exactly offsets its effect as a growth rate under Q of the funds' market values.

4.2 Two-fund contracts with guarantee

Now we consider a version where the benefit payable at time t includes a guarantee,

$$C_t^8 = \max[S_t^1, S_t^2, G_t]. \tag{35}$$

This contract has the advantage that it protects the insured if the values of both funds should be low.

Let $f(z_1, z_2, \rho)$ be the standard normal bivariate density function of Z_1 and Z_2, such that $Z_1 \sim N(0, 1)$, $Z_2 \sim N(0, 1)$, and $\mathrm{Cov}(Z_1, Z_2) = \rho$. We define the binormal distribution function

$$\Phi_2(a, b, \rho) = \int_{-\infty}^{a} \int_{-\infty}^{b} f(z_1, z_2, \rho) dz_1 dz_2$$

—that is, as the volume under the probability density function bounded above by a and b.

The results for the remaining contracts will be expressed in terms of the function Φ_2. See, e.g., Hull [1993, p. 245] on how to evaluate Φ_2 numerically, and Abramowitz and Stegun [1970] for a more extensive discussion.

Lemma 4.2. *The market value at time zero for the benefit* $C_t^8 = \max[S_t^1, S_t^2, G_t]$ *is*

$$V_t^8 = S_0^1 \Phi_2(\beta_1^t, \alpha_1^t, \rho_x) + S_0^2 \Phi_2(-\beta_2^t, \alpha_2^t, -\rho_y) + G_t e^{-rt} \Phi_2(-\gamma_1^t, -\gamma_2^t, \hat{\rho}), \quad (36)$$

where

$$\rho_x := \frac{1}{\Psi} \left(V_x - \frac{\Gamma}{V_x} \right),$$

$$\rho_y := \frac{1}{\Psi} \left(\frac{\Gamma}{V_y} - V_y \right),$$

$$\hat{\rho} := \frac{\Gamma}{V_x V_y},$$

$$\alpha_1^t := \frac{1}{V_x \sqrt{t}} \left[\ln \left(\frac{S_0^1}{G_t} \right) + \left(r + \frac{1}{2} V_x^2 \right) t \right],$$

$$\alpha_2^t := \frac{1}{V_y \sqrt{t}} \left[\ln \left(\frac{S_0^2}{G_t} \right) + \left(r + \frac{1}{2} V_y^2 \right) t \right],$$

$$\gamma_1^t := \frac{1}{V_x \sqrt{t}} \left[\ln \left(\frac{S_0^1}{G_t} \right) + \left(r - \frac{1}{2} V_x^2 \right) t \right], \quad (37)$$

and

$$\gamma_2^t := \frac{1}{V_y \sqrt{t}} \left[\ln \left(\frac{S_0^2}{G_t} \right) + \left(r - \frac{1}{2} V_y^2 \right) t \right]. \qquad \square$$

4.3. Two-fund contracts with cap

In this paragraph the benefit is similar to the previous one, but a cap has replaced the guarantee. We consider the benefit payable at time t,

$$C_t^9 = \min[\max[S_t^1, S_t^2], K_t]. \quad (38)$$

Lemma 4.3. *The market price at time zero for the benefit* $C_t^9 = \min[\max[S_t^1, S_t^2], K_t]$ *is*

$$V_t^9 = S_0^1\Phi_2(\beta_1^t, -\alpha_3^t, -\rho_x) + S_0^2\Phi_2(-\beta_2^t, -\alpha_4^t, \rho_y) + K_t e^{-rt}[1 - \Phi_2(-\gamma_3^t, -\gamma_4^t, \hat{\rho})], \tag{39}$$

where

$$\alpha_3^t := \frac{1}{V_x\sqrt{t}}\left[\ln\left(\frac{S_0^1}{K_t}\right) + \left(r + \frac{1}{2}V_x^2\right)t\right],$$

$$\alpha_4^t := \frac{1}{V_y\sqrt{t}}\left[\ln\left(\frac{S_0^2}{K_t}\right) + \left(r + \frac{1}{2}V_y^2\right)t\right],$$

$$\gamma_3^t := \frac{1}{V_x\sqrt{t}}\left[\ln\left(\frac{S_0^1}{K_t}\right) + \left(r - \frac{1}{2}V_x^2\right)t\right],$$

and

$$\gamma_4^t := \frac{1}{V_y\sqrt{t}}\left[\ln\left(\frac{S_0^2}{K_t}\right) + \left(r - \frac{1}{2}V_y^2\right)t\right]. \qquad \square$$

4.4. Two-fund contracts with cap and guarantee

The benefit is now a combination of the two previous ones, such that both a cap and guarantee are included. Thus we consider the benefit

$$C_t^{10} = \min[\max[S_t^1, S_t^2, G_t], K_t]. \tag{40}$$

This benefit can be split into two components as

$$C_t^{10} = \max[S_t^1, S_t^2, G_t] - \max[S_t^1 - K_t, S_t^2 - K_t, 0].$$

The first term is identical to a simple two-fund contract with guarantee, returning the maximum of the two funds and the guaranteed amount, The second term subtracts any positive excess value of any of the funds above the cap.

Lemma 4.4. *The market price at time zero for the benefit* $C_t^{10} = \min[\max[S_t^1, S_t^2, G_t], K_t]$ *is*

$$V_t^{10} = S_0^1[\Phi_2(\beta_1^t, \alpha_1^t, \rho_x) - \Phi_2(\beta_1^t, \alpha_3^t, \rho_x)] + S_0^2[\Phi_2(-\beta_2^t, \alpha_2^t, -\rho_y) - \Phi_2(-\beta_2^t, \alpha_4^t, -\rho_y)$$

$$+ G_t e^{-rt}\Phi_2(-\gamma_1^t, -\gamma_2^t, \hat{\rho}) + K_t e^{-rt}[1 - \Phi_2(-\gamma_3^t, -\gamma_4^t, \hat{\rho})]. \tag{41}\square$$

Observe that V_t^9 follows from V_t^{10} by letting $G_t := 0$. Similarly, notice that the expression for V_t^8 can be obtained from V_t^{10} by letting K_t tend to infinity.

5. Flexible unit-linked contracts

Subsection 2.6 introduced flexible contracts, where at some fixed, future point in time τ the link characteristics, if any, for the remaining term of the contract are irreversibly determined. Recall that from time τ onward, the benefit is set equal to one of the univariate processes belonging to the set $\{S_t^I, D_t^J\}$. The resulting insurance benefit is $C_t = C_t(\Theta_\tau(S_t^I, D_t^J | \mathfrak{F}_\tau))$ for some choice function $\Theta_\tau(\cdot | \mathfrak{F}_\tau)$, conditional on the information set $\mathfrak{F}_\tau$, for the time interval $[\tau, T]$. It is assumed that the discounted benefit process $\{e^{-r(t-\tau)}C_t: t \geq \tau\}$ is a martingale with respect to the probability measure Q and the filtration $\{\mathfrak{F}_\tau: t \geq \tau\}$. This condition is automatically satisfied, when linking to a single mutual fund. The condition is also satisfied for traditional, non-unit-linked benefits, without reference to any particular probability measure or filtration, provided the benefits over the time interval $[\tau, T]$ are of the form $D_t = G_t = G_0 e^{rt}$ for a guarantee scheme, and $D_t = K_t = K_0 e^{rt}$ for a cap.

Below we present a number of alternative specifications of the set $\{S_t^I, D_t^J\}$, from which one and only one element will be selected at time τ. We also assume that the choice function $\Theta_\tau(\cdot | \mathfrak{F}_\tau)$ is based on market values or, equivalently, premium reserves, of the insurance contract at the link selection time. Heuristically, one may interpret the choice function as the insured wanting to maximize the market value at time τ of the insurance contract, given a particular alternative elementary benefit pattern, whereas the insurer wants to minimize the corresponding time τ market value of the contract.

Assuming the insured is alive at time τ, the survival probability ${}_{T-\tau}p_{x+\tau}$ replaces ${}_Tp_x$ as the relevant survival probability. In analogy with the comments following Proposition 2.2, for the pure endowment contracts, the time τ market values are $\pi_\tau(S^i) = {}_{T-\tau}p_{x+\tau}S_\tau^i$ for a standard unit-linked contract and $\pi_\tau(D^j) = {}_{T-\tau}p_{x+\tau}D_\tau^j$ for a traditional contract. For term insurance contracts, the similar time τ market values are $\pi_\tau^1(S^i) = [1 - {}_{T-\tau}p_{x+\tau}]S_\tau^i$ for a standard unit-linked contract and $\pi_\tau^1(D^j) = [1 - {}_{T-\tau}p_{x+\tau}]D_\tau^j$ for a traditional contract but with deterministic benefits escalating at the riskfree rate.

In computing the time τ market values or premium reserves for either type of insurance contract the probability terms ${}_{T-\tau}p_{x+\tau}$ and $[1 - {}_{T-\tau}p_{x+\tau}]$ factor out from the time τ values of the individual benefits. Thus, to compare the market values of the contracts at time τ, it suffices to compare the market prices S_τ^i of a the mutual fund of a standard, unit linked contract, and the discounted amount D_τ^j associated with a traditional contract.

As a further justification for assuming that the choice function $\Theta_\tau(\cdot | \mathfrak{F}_\tau)$ is based on market values of the insurance contract at the link selection time, suppose the insured can choose exactly one benefit C_t from the menu $\{S_t^I, D_t^J\}$ at the link selection date τ. Also assume that the contract based on the benefit C_t^* has the highest market value π_τ^* and thus the highest benefit value C_τ^*. If, for some reason, the insured does not choose the contract with the highest market value but rather one based on the benefit C_t' with current benefit value C_τ' and the lower current contract value π_τ', he receives a cash refund at time τ of the market value loss $[\pi_\tau^* - \pi_\tau']$. Alternatively, the benefit C_t' is increased by the factor $\pi_\tau^*/\pi_\tau' > 1$. In either case it can be shown that valuing a contract based on the market value maximizing benefit C_t^* is equivalent to valuing a contract based on the benefit C_t', when the latter contract is supplemented by either a cash refund or a running proportional increase in the benefit, as described above.

For each specification of the set $\{S_t^I, D_t^J\}$, imposing the martingale condition on the elementary benefit functions, and assuming a market-value-based choice function, separate lemmas state the derived time-zero market value $V_\tau^* = E^Q[e^{-r\tau} C_\tau]$ of C_t for all $t \geq \tau$. Insertion of these lemmas into Proposition 2.2 then gives the single premiums at time zero of the various flexible unit-linked contracts discussed below.

5.1. Flexible one-fund unit-linked contract with conversion option

The benefit of this contract is

$$C_t^{11} = \begin{cases} C_t^A \text{ if } t < \tau, \\ S_t \text{ if } S_\tau \geq G_\tau \text{ and } t \geq \tau, \\ G_t \text{ if } G_\tau > S_\tau \text{ and } t \geq \tau. \end{cases} \tag{42}$$

This contract can be interpreted as giving the insured the right to replace the standard, simple unit-linked benefit with the certain benefit of a traditional insurance. Optimal exercise of this conversion option has been built into the contract.

Lemma 5.1. *The time zero market value of the benefit* (42) *from a flexible one-fund unit-linked contract with conversion option is*

$$V_\tau^{*11} = S_0\Phi(d_1^\tau) + G_0\Phi(d_2^\tau), \tag{43}$$

where

$$d_1^\tau := \frac{1}{\sigma\sqrt{\tau}}\left[\ln\left(\frac{S_0}{G_0}\right) + \frac{1}{2}\sigma^2\tau\right]$$

and

$$d_2^\tau := \frac{1}{\sigma\sqrt{\tau}}\left[\ln\left(\frac{S_0}{G_0}\right) - \frac{1}{2}\sigma^2\tau\right]. \qquad \square$$

Comparing the arguments d_1^τ and d_2^τ with their d_1^t and d_2^t counterparts from equations (23), observe that the term rt has ben dropped, as the time-zero certain benefit from G_0 has replaced the time t certain benefit G_t, because of the martingale assumption. Without further explicit statements, similar formulation changes will apply to the arguments d_3^τ, d_4^τ, α_1^τ, α_2^τ, α_3^τ, α_4^τ, γ_1^τ, γ_2^τ, γ_3^τ, and γ_4^τ as well because of the martingale assumptions with respect to G_t and K_t.

5.2. Flexible one-fund unit-linked contract with buy-back option

For this contract the benefit is

$$C_t^{12} = \begin{cases} C_t^A \text{ if } t < \tau, \\ S_t \ \text{ if } S_\tau \leq K_\tau \text{ and } t \geq \tau, \\ K_t \ \text{ if } K_\tau < S_\tau \text{ and } t \geq \tau. \end{cases} \tag{44}$$

This contract can be interpreted as the insurer optimally exercising his right to replace the standard, simple unit-linked benefit with the traditional insurance certain benefit of a lesser market value at the link selection time τ.

Lemma 5.2. *The time-zero market value of the benefit* (44) *from a flexible one-fund unit-linked contract with buy-back option is*

$$V_\tau^{*12} = S_0\Phi(-d_3^\tau) + K_0\Phi(d_4^\tau). \tag{45}\square$$

5.3. Flexible one-fund unit-linked contract with conversion option and buy-back option

This contract is a combination of the two previous ones. The benefit is

$$C_t^{13} = \begin{cases} C_t^A \text{ if } t < \tau, \\ S_t \ \text{ if } S_\tau \leq K_\tau \text{ and } S_\tau \geq G_\tau \text{ and } t \geq \tau, \\ K_t \ \text{ if } K_\tau < S_\tau \text{ and } t \geq \tau, \\ G_t \ \text{ if } G_\tau > S_\tau \text{ and } t \geq \tau. \end{cases} \tag{46}$$

Under this contract, both the insured and the insurer have the right to replace the standard, simple unit-linked benefit with a traditional insurance certain benefit, in case of an unfavorable market value of the unit-linked insurance at the decision selection time τ.

Lemma 5.3. *The time-zero market value of the benefit* (46) *from a flexible one-fund unit-linked contract with both conversion option and buy-back option is*

$$V_\tau^{*13} = S_0[\Phi(d_1^\tau) - \Phi(d_3^\tau)] + G_0\Phi(-d_2^\tau) + K_0\Phi(d_4^\tau). \tag{47}$$

5.4. Flexible two-fund switch-link option

The remaining insurance contracts allow the insured to switch link unit by unit to another mutual fund. Assuming the mutual fund with the highest market value at the switch time τ will be selected for linkage from time τ onward, the benefit of the next contract is.

$$C_t^{14} = \begin{cases} C^A \text{ for } t < \tau, \\ S_t^1 \text{ if } S_\tau^1 \geq S_\tau^2 \text{ and } t \geq \tau, \\ S_t^2 \text{ if } S_\tau^2 > S_\tau^1 \text{ and } t \geq \tau. \end{cases} \tag{48}$$

Lemma 5.4. *The time-zero market value of the benefit* (48) *from a flexible contract with two-fund switch-link option is*

$$V_\tau^{*14} = S_0^1 \Phi(\beta_1^\tau) + S_0^2 \Phi(-\beta_2^\tau). \tag{49}\square$$

5.5. Flexible two-fund switch-link option with conversion option

Under this contract, the insured has the right at time τ to choose whether the insurance benefit should be linked to which of the two mutual funds, if any. The choice may also be conversion to a traditional insurance with fixed benefit schedule. The benefit is now given as

$$C_i^{15} = \begin{cases} C_i^A \text{ if } t < \tau \\ S_t^1 \text{ if } S_\tau^1 \geq S_\tau^2 \text{ and } S_\tau^1 \geq G_\tau, \text{ and } t \geq \tau, \\ S_t^2 \text{ if } S_\tau^2 > S_\tau^1 \text{ and } S_\tau^2 \geq G_\tau, \text{ and } t \geq \tau, \\ G_t \text{ if } G_\tau > S_\tau^1 \text{ and } G_\tau > S_\tau^2, \text{ and } t \geq \tau. \end{cases} \tag{50}$$

Lemma 5.5. *The time-zero market value for the benefit (50) from a flexible two-fund switch-link option with conversion option is*

$$V_\tau^{*15} = S_0^1 \Phi_2(\beta_1^\tau, \alpha_1^\tau, \rho_x) + S_0^2 \Phi_2(-\beta_2^\tau, \alpha_2^\tau, -\rho_y) + G_0 \Phi_2(-\gamma_1^\tau, -\gamma_2^\tau, \hat{\rho}). \tag{51}\square$$

5.6. Flexible two-fund switch-link option with buy-back option

The insured may still choose either one of the two mutual funds for linkage, but the insurer may buy back the contract by substituting a traditional insurance with fixed-benefit pattern over time. The resulting benefit can be written as

$$C_t^{16} = \begin{cases} C_t^A \text{ if } t < \tau, \\ S_t^1 \text{ if } S_\tau^1 \geq S_\tau^2 \text{ and } S_\tau^1 \leq K_\tau, \text{ and } t \geq \tau, \\ S_t^2 \text{ if } S_\tau^2 > S_\tau^1 \text{ and } S_\tau^2 \leq K_\tau, \text{ and } t \geq \tau, \\ K_t \text{ if } K_\tau < S_\tau^1 \text{ or } K_\tau < S_\tau{}^2, \text{ and } t \geq \tau. \end{cases} \tag{52}$$

Lemma 5.6. *The time-zero market value of the benefit* (52) *from a flexible two-fund switch-link option with buy-back option is*

$$V_\tau^{*16} = S_0^1 \Phi_2(\beta_1^\tau, -\alpha_3^\tau, -\rho_x) + S_0^2 \Phi_2(-\beta_2^\tau, -\alpha_4^\tau, \rho_y) + K_0[1 - \Phi_2(-\gamma_3^\tau, -\gamma_4^\tau, \hat{\rho})]. \tag{53}$$

$\square$

5.7. Flexible two-fund switch-link option with conversion option and buy-back option

The final contract considered is a combination of the two previous ones. The insured may choose either a mutual fund for linkage or a traditional benefit, whereas the insurer may choose another but higher deterministic benefit by buying back the unit-linked contract preferred by the insured. Thus, in this case the choice function determines the benefit

$$C_t^{17} = \begin{cases} C_t^A \text{ if } t < \tau, \\ S_t^1 \text{ if } S_\tau^1 \geq S_\tau^2 \text{ and } S_\tau^1 \geq G_\tau \text{ and } S_\tau^1 \leq K_\tau, \text{ and } t \geq \tau, \\ S_\tau^2 \text{ if } S_\tau^2 > S_\tau^1 \text{ and } S_\tau^2 \geq G_\tau \text{ and } S_\tau^2 \leq K_\tau, \text{ and } t \geq \tau, \\ K_t \text{ if } K_\tau < S_\tau^1 \text{ or } K_\tau < S_\tau^2, \text{ and } t \geq \tau, \\ G_t \text{ if } G_\tau > S_\tau^1 \text{ and } G_\tau > S_\tau^2, \text{ and } t \geq \tau. \end{cases} \tag{54}$$

Lemma 5.7. *The time-zero market value of the benefit* (54) *from a flexible two-fund switch-link option with conversion option and buy-back option is*

$$V_\tau^{*17} = S_0^1[\Phi_2(\beta_1^\tau, \alpha_1^\tau, \rho_x)] - \Phi_2(\beta_1^\tau, \alpha_3^\tau, \rho_x) + S_0^2[\Phi_2(-\beta_2^\tau, \alpha_2^\tau, -\rho_y) - \Phi_2(-\beta_2^\tau, \alpha_4^\tau, -\rho_y)]$$
$$+ G_0\Phi_2(-\gamma_1^\tau, -\gamma_2^\tau, \hat{\rho}) + K_0[1 - \Phi_2(-\gamma_3^\tau, -\gamma_4^\tau, \hat{\rho})]. \tag{55}\square$$

6. Concluding remarks

Casual observations indicate that unit-linked products may catch a substantial market share. Just as exotic options have gained popularity in financial risk management, exotic unit-linked contracts may become widely used insurance instruments among professionals.

This article presents several new unit-linked insurance contracts with substantial potential for real-life applications.

Compared to usual unit-linked products, these contracts offer added flexibility and/or altered exposure to financial risk for either the insurer or the insured. The article shows calculations of single premiums of a variety of pure endowment and term insurance contracts. The results are derived using a somewhat idealized economic model based on the reasonable principle of no-arbitrage opportunities, equivalent to the discounted price processes being martingales.

The single premiums of the contracts are computed as expectations under a risk-adjusted probability measure. Risk neutrality with respect to mortality can be defended by referring to the law of large numbers, assuming the company has a large number of independent mortality risks. Risk neutrality with respect to financial risk is not a reasonable assumption. Increasing the number of identical contracts does not eliminate the financial risk exposure. A crucial assumption is the independence between mortality risk and financial risk.

An issue we have not addressed is how the insurance company can reduce the associated financial risk by employing so-called hedging strategies. The benefits we consider are log-normally distributed and therefore of finite variance. As briefly mentioned in Section 2, our model is dynamically complete. These facts imply that there exist trading strategies that duplicate the payoffs of the benefits. For insurance contracts these duplicating strategies will depend on mortality factors and will generally not be self-financing, see Aase and Persson [1994]. One may expect that the popularity of unit-linked life insurance products to some extent will depend on how successfully these theoretical strategies may be adapted to operational portfolio strategies applicable in real financial markets.

In our model the interest rate is deterministic. Interest-rate risk is another major source of financial uncertainty for the insurance industry. We expect that the results of this article can be generalized to include stochastic interest rates as indicated in Baccinello and Ortu [1993b, 1994, 1995], Nielsen [1993], Nielsen and Sandmann [1995], and Persson [1994].

The various unit-linked contracts discussed may indeed seem exotic to many insurance economists. However, in some more traditional insurance contracts the benefit is nominally fixed, but a proportion of the company's profit is customarily added on to the benefit. Suppose, in addition, that the company maintains a policy of imposing an upper limit on the benefit. Then the insurance contract informally resembles the one studied in Subsection 3.4—that is, the single-fund capped benefits with a guarantee. One conspicuous difference is that the "mutual fund' in the more traditional context corresponds to the insurance company itself (disregarding the difference between profit and market value). However, the insurance customer may very well prefer to have the benefit linked to some other financial asset, rather than to the insurance company's profit. Moral hazard is less of a problem with "true" unit-linked contracts.

A more important advantage of unit-linked products, however, is the possibility for influencing the financial risk of the policy by choosing both the form of the contract and the financial asset to which the benefits are linked. Both understanding of the features of various contracts as well as proper valuation and premium assessments are required for a sustainable development of the unit-linked products.

Appendix

Proof of Lemma 3.2

The benefit of this contract can be written as $\max[S_t - G_t, 0] + G_t$. The market price at time zero of the second term is $G_t e^{-rt}$. Black and Scholes [1973] value the first term as a European call option and obtain the well-known expression $S_0\Phi(d_1^t) - G_t e^{-rt}\Phi(d_2^t)$ as its market value at time zero.

Alternatively, the benefit can be written as $S_t + \max[G_t - S_t, 0]$. The time zero value of S_t is S_0, as the fund does not pay any dividends. The second term is recognized as the payoff to a European put option, with its Black and Scholes market value of $G_t e^{-rt}\Phi(-d_2^t) - S_0\Phi(-d_1^t)$.

In either case, the market value of the claim $\max[S_t, G_t]$ is then obtained by adding the time-zero market values of the two components, using the market value additivity property as well as the symmetry property $\Phi(d) = 1 - \Phi(-d)$ for evaluation of a standard normal random variable at any value d. □

Proof of Lemma 3.3

Immediate from either of the three reformulations of the benefit, combined with value additivity and symmetry. E.g., from $C_t^3 = (S_t + K_t) - \max[S_t, K_t]$ and Lemma 3.2, $V_t^3 = S_0 + K_t e^{-rt} - [S_0\Phi(d_3^t) + K_t e^{-rt}\Phi(-d_4^t)] = S_0[1 - \Phi(d_3^t)] + K_t e^{-rt}[1 - \Phi(-d_4^t)]$, which yields the result. □

Proof of Lemma 3.4

The benefit can be rewritten in a number of ways, including $C_t^4 = G_t + \max[S_t - G_t, 0] - \max[S_t - K_t, 0]$. That is, C_t^4 equals the guarantee amount plus a call option with striking price G_t minus a call option with striking price K_t. Thus, $V_t^4 = G_t e^{-rt} + [S_0\Phi(d_1^t) - G_t e^{-rt}\Phi(d_2^t)] - [S_0\Phi(d_3^t) - K_t e^{-rt}\Phi(d_4^t)]$, by value additivity. Collection of terms and symmetry yield the stated expression. □

Proof of Proposition 3.1

From Proposition 2.1 and the events of death and survival being mutually exclusive, the single premium $\pi_5^* = {}_Tp_x E^Q[v(T)S_T] + (1 - {}_Tp_x)E^Q[v(T)G_T]$. Using Lemma 3.1, the constant interest rate in the discount function, and G_T being nonstochastic, this expression simplifies into the stated one. □

Proof of Proposition 3.2

By value additivity and independence between mortality and financial risk, $\pi_6^* = {}_Tp_x E^Q[v(T)\max[S_T, G_T]] + (1 - {}_Tp_x)E^Q[v(T)\min[S_T, G_T]]$. From Lemmas 3.2 and 3.3, using $G_T = K_T$ implying $d_1^T = d_3^T$ and $d_2^T = d_4^T$, $\pi_6^* = {}_Tp_x[S_0\Phi(d_1^T) + G_T e^{-rT}\Phi(d_2^T)] + (1 - {}_Tp_x)[S_0\Phi(-d_1^T) + G_T e^{-rT}\Phi(d_2^T)]$. The result follows from the symmetry property of normal distributions and collections of terms. □

Proof of Proposition 3.3

The result follows directly by substitution of the expressions for π_6^* and π_5^*, collection of terms, and the symmetry of the normal distribution. □

Proof of Lemma 4.1

The payoff of this contract can be written as $\max[S_t^1 - S_t^2, 0] + S_t^2$. The market price at time zero of the last term is $S_0{}^2$. Margrabe [1978] values the first term using a different but equivalent methodology and obtains in our notations the expression $S_0^1\Phi(\beta_1^t) - S_0^2\Phi(\beta_2^t)$ as its market value at time zero. The result then follows by the market value additivity property. □

Proof of Lemma 4.2

Stulz [1982] values the claim $\max[[\min(S_t^1, S_t^1) - G_t], 0]$ in a similar setting using a different but equivalent methodology. Rewritten using our notation, his formula for the market value at time zero is

$$S_0^1\Phi_2(-\beta_1^t, \alpha_1^t, -\rho_x) + S_0^2\Phi_2(\beta_2^t, \alpha_2^t, \rho_y) - G_t e^{-rt}\Phi_2(\gamma_1^t, \gamma_2^t, \hat{\rho}).$$

(In case of more than two funds, note that Johnson [1987] extends Stulz's result to the maximum or minimum of several assets, getting results involving the multivariate normal distribution function.)

Now observe that $\max[S_t^1, S_t^2, G_t] =$

$$\max[S_t^1 - G_t, 0] + \max[S_t^2 - G_t, 0] - \max[[\min(S_t^1, S_t^2) - G_t], 0] + G_t.$$

The market value at time zero of the first terms are obtained by the Black and Scholes European call option formula, the market value of the third term follows from the above formula of Stulz, and the market value of the last term is $G_t e^{-rt}$. The result now follows by using the market value additivity property and the standard results $\Phi_2(a, b, \rho) = \Phi(a) - \Phi_2(a, -b, -\rho)$ and $\Phi_2(a, b, \rho) = \Phi(a) + \Phi(b) - 1 + \Phi_2(-a, -b, \rho)$. □

Proof of Lemma 4.3

Observe that this benefit can be rewritten as $C_t^9 = \max[S_t^1, S_t^2] + K_t - \max[S_t^1, S_t^2, K_t]$. The market value at time zero of the first term is given by Lemma 4.1, and the market value of the second term is $K_t e^{-rt}$. The market value of the last term is given by Lemma 4.2, by an expression corresponding to the one for V_t^8, but where α_3^t, α_4^t, γ_3^t, and γ_4^t replace α_1^t, α_2^t, γ_1^t, and γ_2^t, respectively. The result is obtained by collecting terms, using the standard results mentioned in the last sentence of the proof of Lemma 4.2. □

Proof of Lemma 4.4

Observe that $\max[S_t^1 - K_t, S_t^2 - K_t, 0] = \max[S_t^1, S_t^2, K_t] - K_t$. Thus, we rewrite $C_t^{10} = \max[S_t^1, S_t^2, G_t] + K_t - \max[S_t^1, S_t^2, K_t]$. The market values at time zero of the first and third terms follow by Lemma 4.2. The market value of the second term is $e^{-rt}K_t$. The result then follows by standard properties of binormal distributions.

Proof of Lemma 5.1

Consider the following equalities:

$$V_\tau^* = E^Q[e^{-rt}C_t] = E^Q[e^{-r\tau}E^Q[e^{-r(t-\tau)}C_t \mid \mathfrak{F}_\tau]] = E^Q[e^{-r\tau}C_\tau] \text{ for } \tau \leq t.$$

Substitution from the benefit function shows that the last equation is equivalent to $E^Q[e^{-r\tau}\max[S_\tau, G_\tau]]$. The result then follows from Lemma 3.2, with t replaced by τ. □

Proof of Lemmas 5.2–5.7

Mimicking the proof of Lemma 5.1, the results follow by calling on Lemmas 3.3, 3.4, 4.1, 4.2, 4.3, and 4.4, respectively. □

Acknowledgments

The authors are grateful to an anonymous referee and to the editors of this special issue for helpful comments and constructive suggestions. The usual disclaimers apply.

References

AASE, K.K., and PERSSON, S.A. [1994]: "Pricing of Unit-Linked Life Insurance Policies," *Scandinavian Actuarial Journal*, 26–52.

ABRAMOWITZ, M., and STEGUN, I.A. (Eds.). [1970]: *Handbook of Mathematical Functions*, National Bureau of Standards Applied Mathematics Series 55, U.S. Government Priting Office, Washington, D.C. 20402.

BACINELLO, A.R., and ORTU, F. [1993a]: "Pricing Equity-Linked Life Insurance with Endogenous Minimum Guarantees," *Insurance: Mathematics and Economics*, 12, 245–257.

BACINELLO, A.R., and ORTU, F. [1993b]: "Pricing Guaranteed Securities-Linked Life Insurance Under Interest-Rate Risk," *Actuarial Approach for Financial Risk*, Transactions of the Third AFIR International Colloquium, Rome, 35–55.

BACINELLO, A.R., and ORTU, F. [1994]: "Single and Periodic Premiums for Guaranteed Equity-Linked Life Insurance under Interest Rate Risk: The Lognormal + Vacisek Case," *Financial Modelling*, L. Peccati and M. Viren (Eds.), Physica-Verlag, Heidelberg, Germany, 1–25.

BACINELLO, A.R., and ORTU, F. [1995]: "Fixed Income Linked Life Insurance Policies with Minimum Guarantees: Pricing Models and Numerical Results," Working paper, University of Cassino, Via Mazzaropi, 03043 Cassino (FR), Italy.

BLACK, F., and SCHOLES, M. [1973]: "The Pricing of Options and Corporate Liabilities," *Journal of Political Economy*, 81, 637–654.

BOYLE, P.P., and SCHWARTZ, E.S. [1977]: "Equilibrium Prices of Guarantees Under Equity-Linked Contracts," *Journal of Risk and Insurance*, 44, 639–680.

BRENNAN, M.J., and SCHWARTZ, E.S. [1976]: "The Pricing of Equity-Linked Life Insurance Policies with an Asset Value Guarantee," *Journal of Financial Economics*, 3, 195–213.

BRENNAN, M.J., and SCHWARTZ, E.S. [1979a]: "Alternative Investment Strategies for the Issuers of Equity Linked Life Insurance with an Asset Value Guarantee," *Journal of Business*, 52, 63–93.

BRENNAN, M.J., and SCHWARTZ, E.S. [1979b]: *Pricing and Investment Strategies for Equity-Linked Life Insurance*, Monograph no. 7/The S.S. Huebner Foundation for Insurance Education, Wharton School, University of Pennsylvania, Philadelphia.

BROADIE, M., and DETEMPLE, J. [1995]: "American Capped Call Options on Dividend-Paying Assets," *The Review of Financial Studies*, 8, 161–191.

DELBAEN, F. [1990]: "Equity Linked Policies," *Bulletin Association des Actuaires Belges*, 33–52.

DUFFIE, D. [1992]: *Dynamic Asset Pricing Theory*, University Press, Princeton, New Jersey.

DYBVIG, P.H., and ROSS, S.A. [1992]: "Arbitrage," in *The New Palgrave Dictionary of Money and Finance*, Peter Newman, Murray Milgate, and John Eatwell (Eds.), Macmillan Press, London.

HARRISON, J.M., and KREPS, J. [1979]: "Martingales and Multiperiod Securities Markets," *Journal of Economic Theory*, 20, 381–408.

HULL, J.C. [1993]: *Options, Futures, and Other Derivative Securities, Second Ed.*, Prentice-Hall International, Englewood Cliffs, NJ 07632.

JOHNSON, H. [1987]: "Options on the Maximum or the Minimum of Several Assets," *Journal of Financial and Quantitative Analysis*, 22, 277–283.

MARGRABE, W. [1978]: "The Value of an Option to Exchange One Asset for Another," *Journal of Finance*, 33, 177–186.

NIELSEN, J.A. [1993]: "Equity-Linked Life Insurance Contracts in an Economy with a Stochastic Development of the Term Structure of Interest Rates," Department of Operations Research, Aarhus University.

NIELSEN, J.A., and SANDMANN, K. [1995]: "Equity-Linked Life Insurance: A Model with Stochastic Interest Rates," *Insurance: Mathematics and Economics*, 16, 225–253.

PERSSON, S.A. [1994]: "Pricing Life Insurance Contracts Under Financial Uncertainty," Doctoral dissertation, Norwegian School of Economics and Business Administration, Bergen.

STULZ, R. [1982]: "Options on the Minimum or Maximum of Two Assets," *Journal of Financial Economics*, 10, 161–185.

The Geneva Papers on Risk and Insurance Theory, 21: 65–102 (1996)

Uniqueness of the Fair Premium for Equity-Linked Life Insurance Contracts

J. AASE NIELSEN
Department of Operations Research, University of Aarhus, Bldg. 530, Ny Munkegade, DK-8000 Aarhus C, Denmark
atsjan@mi.aau.dk

KLAUS SANDMANN
Department of Statistics, Rheinische Friedrich-Wilhelms-Universität Bonn, Adenauerallee 24-42, D-53113 Bonn, Germany
sandmann@addi.or.uni-bonn.de

Abstract

An equity-linked life insurance contract combines an endowment life insurance and an investment strategy with a minimum guarantee. The benefit of this contract is determined by the guaranteed amount plus a bonus equal to a call on the portfolio. This bonus is similar to an Asian option. This article analyzes the relationship between the periodic insurance premium and its proportional share invested into the portfolio. For a general model of the financial risks we show the existence and uniqueness of an insurance premium. Furthermore the premium is strictly increasing and convex as a function of the share invested.

Key words: Asian option, forward risk adjusted measure, life insurance, Monte Carlo simulation, stochastic interest rates

1. Introduction

An equity-linked life insurance contract combines in a particular way an endowment[1] life insurance and a portfolio investment strategy. The life insurance component guarantees in the case of maturity of the contract or the death of the insured his or her heirs a fixed amount of money, the so-called guaranteed amount. This guaranteed amount may be constant for the term of the contract or may be an initially specified function of time. Especially in combination with an investment strategy the insured person may choose a decreasing guaranteed amount in time. In this case the insured or his or her heirs would benefit from a relatively high guaranteed amount in the case of an unexpected death within the first period of the contract. The decrease of this insured guaranteed amount over time would then be compensated by the value of the investment strategy. A variety of different contracts are described in Ekern and Persson [1996]. In particular this idea underlies an equity-linked life insurance. In addition to the life insurance component a proportion of the typically periodic premium is invested into a mutual fund. In the case of death of the insured or at maturity of the contract the benefit is determined by the maximum between the guaranteed amount and the market value of the portfolio. The payoff of the contract equals the guaranteed amount plus a nonnegative bonus. This bonus corresponds to a call option on the portfolio value with an exercise price equal to the guaranteed amount. Via the return of the investment strategy the insured participate in the economic growth of the particular fund. Existing equity-linked life insurance contracts are based on interest rate, equity indexes, or real estate portfolios. Thus the bonus of the insurance contracts is not as such related to the acquisitions of the insurance company. In addition the portfolio value—that is, the financial risk—is insured with a lower value equal to the guaranteed amount. For an equity-linked life insurance contract the exercise of this financial insurance component is conditioned on the death of the insured. The contracts take into consideration the technical life insurance risk as well as financial risks.

In this article we will analyze a contract with periodic premium payments.[2] At each premium date the insurance company invests a certain fraction of the premium in a specific mutual fund. The guaranteed amount will be a function of time and the periodic premium. The bonus has similarities to an Asian option with the underlying asset consisting of what currently has been invested in the fund at the premium payment dates. The exercise price for this Asian-like option is the guaranteed amount. The development in the financial market that consists of the specified fund and bonds will be described by Itô processes. The model for the contract is very flexible: through the choice of the share of the premium invested in the security and the composition of the guaranteed amount, we can create many different profiles for the possible cash flow of insurance contracts. For given contractual conditions we will analyze methods to the determination of the fair premium, which creates equality between the cost of the contract and the market value of benefit.

In the literature single and periodic equity-linked contracts have been analyzed in Brennan and Schwartz [1976, 1979], Delbaen [1990], Aase and Persson [1994], Persson [1994], Bacinello and Ortu [1993a, 1993b, 1994], and Nielsen and Sandmann [1995]. However, it is only in Bacinello and Ortu [1994] and Nielsen and Sandmann [1995] that the theory is extended to cover stochastic interest-rate dynamics in the periodic premium case. Taking into consideration that the usual term of these insurance contracts is equal to twelve to thirty years, it is necessary to incorporate a stochastic behavior of the term structure of interest rates. In Nielsen and Sandmann [1995] we analyzed in detail, applying Monte Carlo simulation, the case where the guaranteed amount was a function of just the periodic premium, and from the simulations some conjectures were raised. These conjectures will in this article be analyzed on a theoretical basis and extended to allow for guaranteed amounts depending as well on the time. The results will be supported by simulation experiments, and concerning the specific Asian option analysis comparisons will be made to statements in the literature on the pricing of such assets.

The content of the paper is as follows. In Section 2 the notation and the specification of the contract we analyze are presented. Section 3 is devoted to an analysis of the stochastic model underlying the equity and bond markets. In Section 4 the fair premium problem is stated, and a number of theorems are presented. In the theorems we develop the interrelationships between the fair premium, the functional specification of the guaranteed amount, and the share of the premium deemed to be invested in the fund at each premium payment date. Then in Section 5 we apply approximation techniques in pricing Asian options and discuss their validity for the insurance situation. Section 6 contains numerical results based on simulations and on the approximation by Vorst [1992]. In Section 7 we finally conclude.

2. Notation and definition of the contract

An equity-linked life insurance contract is an agreement between a buyer and a seller, where the buyer is committed to pay—typically at periodic intervals and until the maturity of the contract or the death of the buyer, whichever comes first—a predetermined premium to the seller. At maturity or death of the buyer, the seller is committed to deliver a payment in accordance to the agreement settled when the contract was written. This payment, the benefit, is the maximum of (1) a function depending on the periodic premium and on the history of the spot price of the underlying mutual fund from the date of settlement to the expiration date of the contract and (2) a nonrandom guaranteed amount also depending on the periodic premium. To precisely define the insurance contract we apply the following notation:

- The value of the *mutual fund* at time t is denoted by $S(t)$. Relative to the value $S(t)$ the payoff of the equity-linked life insurance will be calculated at time t.
- The equity-linked life insurance contract includes the specification of

  K the *periodic premium* paid by the insured,

  a the *share* of the periodic premium invested in a mutual fund, $0 \leq a \leq 1$,

  t_i the set of *premium payment dates*, $i = 0, 1, 2, \ldots, n - 1$; $t_0 = 0$,

  t_n the *maturity date*, $t_n = T$, and

  $g(t, K)$ the *guaranteed amount*. The guaranteed amount is assumed to be a nonstochastic function of the periodic premium K and the time t.
- *The reference portfolio* relative to which the final payoff is calculated is defined as the portfolio obtained by investing an amount $a{\cdot}K$ at each of the dates t_i, $i = 0, 1, 2, \ldots, n - 1$, in the fund with price process $S(t)$. The market value of the reference portfolio at time t_i equals

$$aK \sum_{j=0}^{i} \frac{S(t_i)}{S(t_j)}.$$

- *The benefit* from the insurance contract payed at time t if the contract terminates at this date is therefore determined by

$$g(t, K) + V(t, t, K) = g(t, K) + \left[a \cdot K \cdot \sum_{i=0}^{n^*(t)-1} \frac{S(t)}{S(t_i)} - g(t, K) \right]^+$$

  with $n^*(t) := \min\{j \mid t_j \geq t\}$ and $[X - g(t, K)]^+ := \max\{X - g(t, K), 0\}$.
- The *fair periodic premium* K^* is defined as the periodic premium K for which the value at date t_0 of the benefit equals the value at the same date of the premium payments, where the latter could also be denoted the cost of the insurance contract.
- The *term structure of interest rates* is determined by

  $r(t)$ the instantaneous risk-free rate of interest at time t, and

  $D(t, t')$ the price at date t of a zero coupon bond with maturity date t', $t \leq t'$.

The benefit is the proceeds from a financial contract and its price at time t_0 will be found in accordance to the absence of arbitrage opportunities in the financial market. The share a is one of the parameters that should be negotiated when the contract is settled. The size of aK determines together with the sequence of prices of the mutual fund the value of the reference portfolio through which the obligations of the insurance company are given. However, it should be pointed out that the insurance company has no obligation to invest the amount aK in the mutual fund at each premium payment date. The company[3] will, of course, invest not only aK but K in accordance to a strategy that takes into account the overall interests of the company, and as an extreme situation this could mean that it has no investment in the mutual fund at all.

3. The stochastic model

In order to analyze an equity-linked life insurance contract we have to model three sources of uncertainty that will affect the premium of such a contract. First, we have to consider the death probability. Usually the death and survival distribution is determined by the insurer using historical data. We assume that the death distribution is stochastically independent from the financial risks, which we will discuss in more detail shortly.[4] Furthermore, we assume that the insurer is risk neutral with respect to mortality. Since the death distribution depends on the age of the insured person we implicitly assume that the insurer can perfectly diversify the death uncertainty within each group. For standard endowment life insurance contracts this is a usual and acceptable assumption. Looking at equity-linked life insurance contracts, we have at least to remember that the total amount of contracts is much smaller than those of the standard case, and it is not obvious that the law of large numbers can be applied.[5]

While the death distribution determines the payment date, the two further sources of risk—the price risk of the underlying mutual fund and the interest-rate risk—determine the size of the benefit, if the contract terminates. Moreover, we will argue that these financial risks are responsible for only the bonus part of the insurance contract—that is, the ex ante uncertain payment that the insurer will pay in addition to the guaranteed amount. Clearly, the stochastic development of the underlying mutual fund will not affect the guaranteed amount, but since we have to consider the discounted value of the total payments—that is, a pathwise discounting of the payments—it is not obvious that the interest-rate risk influences only the bonus part of the contract. In order to clarify this we now explicitly model the two financial markets.

Let $(\Omega, \mathbb{F}, P^*)$ be a filtered probability space. For the financial markets a continuous time and complete market framework is assumed. More specifically, the mutual fund and the interest-rate market are given by Itô processes of the form

$$dS(t) = r(t)S(t)dt + \sigma_1(t)S(t)dW_1^*(t) + \sigma_2(t)S(t)dW_2^*(t) \tag{1}$$

$$dD(t, t') = r(t)D(t, t')dt + \sigma(t, t')D(t, t')dW_1^*(t) \ \forall t' \quad \forall t \leq t' \tag{2}$$

where $\{W_1^*(t)\}_t$ and $\{W_2^*(t)\}_t$ are independent one-dimensional Brownian motions under the measure P^*. In order to exclude arbitrage possibilities we have to assume certain regularity conditions. First note that the volatility functions $\sigma_1(t)$ and $\sigma_2(t)$ for the mutual fund and $\sigma(t, t')$ $\forall t'$ $\forall t \leq t'$ are not necessarily deterministic. Repeating the arguments of El Karoui, Lepage, Myneni, Roseau, and Viswanathan [1991] the interest-rate market has to satisfy the following regularity assumption:

Assumption 1. *For any maturity $t' \in [0, T]$ we assume that the volatility process $\{\sigma(t, t')\}_{0\leq t\leq t'}$ of the default-free zero coupon bond $D(t, t')$ satisfies the following conditions:*

i. $\{\sigma(t, t')\}_{0\leq t\leq t'}$ *is a continuous and adapted process with* $\sigma(t', t') = 0$.
ii. The partial derivative with respect to maturity $\partial\sigma(t, t')/\partial t'$ *is a continuous and adapted process* $\forall t \leq t'$.

iii. $E_{P^*}\left[\exp\left\{\frac{1}{2}\int_0^{t'} \sigma(t, t')^2 dt\right\}\right] < \infty.$

iv. $E_{P^*}\left[\left|\frac{\partial\sigma(t, t')}{\partial t'}\right|^2\right]$ *is bounded on* $\{(t, t')|0 \leq t \leq t', t' \in [0, T]\}$.

v. There exists a predictable process $\{A(t)\}_t$ *with* $E_{P^*}[A(t)^2] < \infty$ *and* $E_{P^*}[\int_0^T A(t)^2 dt] < \infty$ *such that*

$$\left|\frac{\partial\sigma}{\partial t'}(t, t' + \delta) - \frac{\partial\sigma}{\partial t'}(t, t')\right| \leq A(t) \cdot \delta \quad \forall t \leq t' \quad \forall\delta > 0.$$

With Assumption 1 the discounted zero coupon bond price processes are martingales under the measure P^*—that is, P^* is the risk-neutral or equivalent martingale measure. The solution of the stochastic differential equation (2) is given by

$$D(t, t') = D(t_0, t') \cdot \exp\left\{\int_0^t r(u)du - \frac{1}{2}\int_0^t \sigma(u, t')^2 du + \int_0^t \sigma(u, t')dW_1^*(u)\right\}. \quad (3)$$

The spot rate process $\{r(t)\}_t$ is determined by the bond price processes—that is,

$$r(t') = -\frac{\partial D(t_0, t')}{\partial t'} + \frac{1}{2}\int_0^{t'}\left(\frac{\partial}{\partial t'}\sigma(u, t')^2\right) du + \int_0^{t'}\left(\frac{\partial}{\partial t'}\sigma(u, t')\right) dW_1^*(u). \quad (4)$$

Assumption 2. *For the volatility processes of the mutual fund we assume:*

i. $\{\sigma_1(t)\}_t$ *and* $\{\sigma_2(t)\}_t$ *are one-dimensional, continous, and adapted processes.*
ii. $E_{P^*}[\exp\{\int_0^t \sigma_1^2(u) + \sigma_2^2(u)du\}] < \infty \quad \forall t \leq T$.
iii. $E_{P^*}[\sigma_1^2(t) + \sigma_2^2(t)]$ *is bounded for* $t \in [0, T]$.
iv. There exists a predictable process $\{A(t)\}_t$ *with* $E_{P^*}[A(t)^2] < \infty$ *and* $E_{P^*}[\int_0^T A(t^2)dt] < \infty$ *such that*

$$|\sigma_1(t+\delta) + \sigma_2(t+\delta) - \sigma_1(t) - \sigma_2(t)| \leq A(t) \cdot \delta \;\forall t \quad \forall \delta > 0.$$

The Assumptions 1 and 2 are sufficient for the combined mutual fund and interest-rate market to be arbitrage free. Uniqueness of the measure P^* implies the completeness of the combined bond and mutual fund market. Furthermore, the arbitrage price of a financial contract is determined as the expected discounted value under the risk-neutral measure P^*. The independency between the Brownian motions $\{W_1^*(t)\}_t$ and $\{W_2^*(t)\}_t$ implies that the volatility process $\{\sigma_S(t)\}_t$ of the mutual fund is given by

$$\sigma_S^2(t) := \sigma_1^2(t) + \sigma_2^2(t), \tag{5}$$

which is not necessarily deterministic. Furthermore $\{\sigma_1(t)\}_t$ and $\{\sigma_2(t)\}_t$ determine the correlation process between the mutual fund and the zero coupon bond $D(t, t')$ as

$$\rho(t) := \frac{\sigma_1(t)}{\sigma_S(t)} \quad \forall t', \forall t \leq t'. \tag{6}$$

Vice versa, for a given specification of the volatility process of the underlying mutual fund and the correlation process $\{\rho(t)\}_t$ with $\rho(t) \in [-1, 1]$ a.s. $\forall t \in [0, T]$ the processes $\{\sigma_1(t)\}_t$ and $\{\sigma_2(t)\}_t$ are determined by

$$\sigma_1(t) := \rho(t) \cdot \sigma_S(t) \quad \text{and} \quad \sigma_2(t) := \sqrt{1 - \rho(t)^2} \cdot \sigma_S(t). \tag{7}$$

Under the risk-neutral measure P^* the pricing of a financial contract is done by calculating its expected discounted value. Since the discounting is stochastic, this implies a pathwise consideration, even if the value of the contract is path independent. As shown by Jamshidian [1989, 1991] and Geman, El Karoui, and Rochet [1995], this can be simplified by a change of measure technique—that is, by using a zero coupon bond with maturity T as a numeraire for the stochastic processes $S(t)$ and $D(t, t')$, $t' \leq T$. By Itô's Lemma we can rewrite the stochastic differential equations (1) and (2):

$$d\left(\frac{S(t)}{D(t,T)}\right) = \frac{S(t)}{D(t,T)}([\sigma_1(t) - \sigma(t,T)]dW_1^T(t) + \sigma_2(t)dW_2^T(t)), \tag{8}$$

$$d\left(\frac{D(t,t')}{D(t,T)}\right) = \frac{D(t,t')}{D(t,T)}[\sigma(t,t') - \sigma(t,T)]dW_1^T(t), \tag{9}$$

where $(dW_1^T, dW_2^T) := ((dW_1^*(t) - \sigma(t,T))dt, dW_2^*(t))$ are standard Brownian motions under the so-called T-forward risk-adjusted measure P^T given by the Radon-Nikodym derivative

$$\frac{dP^T}{dP^*} = \exp\left\{\int_{t_0}^T \sigma(t,T)dW_1^*(t) - \frac{1}{2}\int_{t_0}^T \sigma^2(t,T)dt\right\}. \tag{10}$$

This change of measure technique implies that the arbitrage price of a financial contract with payment at T is equal to the discounted expected value under the P^T measure. In particular, the market value $V(t_0, T, K)$ at time t_0 of the bonus payable at time T equals

$$V(t_0, T, K) := D(t_0, T) \cdot E_{P^T}\left[\left[a \cdot K \sum_{i=0}^{n-1} \frac{S(T)}{S(t_i)} - g(T, K)\right]^+\right]. \tag{11}$$

Furthermore, the solution of the stochastic processes (8) and (9) implies that

$$\frac{S(T)}{S(t)} = \frac{D(t_0, t)}{D(t_0, T)} \cdot \exp\left\{\int_{t_0}^{t} (\sigma(u, t) - \sigma(u, T))dW_1^T(u)\right\} \tag{12}$$

$$\exp\left\{\int_t^T (\sigma_1(t) - \sigma(u, T))dW_1^T(u) + \int_t^T \sigma_2(t)dW_2^T(u)\right\}$$

$$\exp\left\{-\frac{1}{2}\int_{t_0}^{t} (\sigma(u, t) - \sigma(u, T))^2 du - \frac{1}{2}\int_t^T((\sigma_1(t) - \sigma(u, T))^2 + \sigma_2(t)^2)du\right\}.$$

So far we have not restricted the volatility processes of the mutual fund and the term structure of interest rates to be deterministic. If we would do so, the mutual fund according to (1) would be a log-normal diffusion, and the term structure model would belong to the Gaussian framework. As a consequence of this specific structure the ratio $S(T)/S(t)$ would be log-normally distributed under the T-forward risk-adjusted measure. The theoretical results in Section 4 are not restricted to this specific case. But it is this special distributional assumption that allows us in Section 5 to discuss different numerical techniques to approximate the solution of the fair premium problem.

4. The fair premium

Definition 3. *Consider an equity-linked life insurance contract determined by the time to maturity T, the age x of the insured person, his or her death distribution π_x, and the guaranteed amount $g : [0, T] \times I\!R^+ \to I\!R^+$. A periodic premium K^* at dates t_i, $i = 0, \ldots, n-1$ is called a* fair premium *if the expected discounted value of the sum of the periodic premiums under the death distribution π_x is equal to the expected benefit of the contract under the death distribution—that is, if K^* is a solution to*

$$0 = K^* \sum_{i=0}^{n-1} D(t_0, t_i)\left[1 - \int_{t_0}^{t_i} \pi_x(t)dt\right]$$

$$- \int_{t_0}^{T} D(t_0, t)g(t, K^*)\pi_x(t)dt - D(t_0, T)g(T, K^*)\left[1 - \int_{t_0}^{T} \pi_x(t)dt\right]$$

$$- \int_{t_0}^{T} V(t_0, t, K^*)\pi_x(t)dt - V(t_0, T, K^*)\left[1 - \int_{t_0}^{T} \pi_x(t)dt\right], \tag{13}$$

where

$$V(t_0, t, K^*) = D(t_0, t)E_{P^t}\left[\left[aK^* \sum_{i=0}^{n^*(t)-1} \frac{S(t)}{S(t_i)} - g(t, K^*)\right]^+\right]$$

$$n^*(t) = \min\{j \mid t_j \geq t\}. \tag{14}$$

The equations (13) and (14) explain the influence on the fair premium of the stochastic nature of the financial market and the death distribution. The pricing principle expressed in (14) shows that the price of a contingent claim is equal to the discounted expected value of its uncertain cash flow in the future, where the expectation has to be taken with respect to the forward risk-adjusted measure. This pricing principle is valid when the share as well as the bond market is uncertain. Assume now that the bond market had a deterministic development. Then a similar pricing principle would be valid with the expectation taken under the risk-neutral measure, P^*. It means that the influence of a stochastic versus a deterministic developing short-term interest rate can be explained as the difference in pricing obtained through the application of the operators E_{P^t} and E_{P^*}. The difference between the two expected values of the Asian-like option gives us (although in an implicit form) the market price of interest-rate risk. Denote the original/real probability measure by P. Assume again that the bond market is deterministic. Then the pricing principle similar to (14) but with the application of E_P instead of E_{P^*} would lead to prices where all components of market price of risk are neglected. The attention is next turned toward equation (13). It is through (13) that the interplay between the death distribution and the riskiness of the financial market enters the analysis. Had we used (13) and (14) with E_{P^t} replaced by E_P, then the fair premium would have been determined in accordance to the classical principle of equivalence.

The fair premium will in the following sequence of propositions and theorems be analyzed to restrict the interrelationships between the share a, the periodic premium K, and the function of the guaranteed amount $g(t, K)$. First note that the martingale property of the ratio $S(t)/S(u)$ $\forall u \leq t$ under the t-forward risk-adjusted measure P^t implies

$$E_{P^t}\left[\sum_{i=0}^{n^*(t)-1} \frac{S(t)}{S(t_i)}\right] = \sum_{i=0}^{n^*(t)-1} \frac{D(t_0, t_i)}{D(t_0, t)}. \tag{15}$$

Proposition 1. *Let $\pi_x(t)$ be the density function of the death distribution. Then*

$$\int_{t_0}^{T}\left[\sum_{i=0}^{n^*(t)-1} D(t_0, t_i)\right]\pi_x(t)dt + \left[\sum_{i=0}^{n-1} D(t_0, t_i)\right]\left[1 - \int_{t_0}^{T} \pi_x(t)dt\right]$$

$$= \sum_{i=0}^{n-1}\left[D(t_0, t_i)(1 - \int_{t_0}^{t_i} \pi_x(t)dt)\right].$$

Proof:

$$\int_{t_0}^{T}\left[\sum_{i=0}^{n^*(t)-1} D(t_0, t_i)\right]\pi_x(t)dt = \sum_{i=0}^{n-1}\int_{t_i}^{t_{i+1}}\left[\sum_{j=0}^{\text{i}} D(t_0, t_j)\right]\pi_x(t)dt$$

$$= \sum_{i=0}^{n-1} D(t_0, t_i)\int_{t_i}^{T}\pi_x(t)dt = \sum_{i=0}^{n-1} D(t_0, t_i)\left[(1 - \int_{t_0}^{t_i}\pi_x(t)dt) - (1 - \int_{t_0}^{T}\pi_x(t)dt)\right]$$

□

Proposition 1 and equation (15) imply that the fair premium in the case of a guaranteed amount $g(t, K) = 0 \; \forall t \forall K$ is the solution of

$$0 = (1 - a)K^* \sum_{i=0}^{n-1} D(t_0, t_i)\left[1 - \int_{t_0}^{t_i}\pi_x(t)dt\right]$$

—that is, $\forall a \in [0, 1]$ the value $K^* = 0$ is the unique fair premium, and for $a = 1$ any K is a fair premium in this situation. Let us now consider conditions on the functional form of a guaranteed amount that are sufficient for the existence of a unique solution K^* for the fair premium problem defined by equation (13) if the guaranteed amount is not identical to zero. In Nielsen and Sandmann [1995] we have derived the existence of a unique fair premium in the case of a time-independent guaranteed amount $g(K)$ and a specific duration type Gaussian term structure model. In the general model for the financial markets we can now extend this to the multiplicative separable time dependent case—that is, where

$$g(t, K) = F(t) \cdot h(K). \tag{16}$$

In the remaining part of this article it is assumed that $g(t, K)$ can be separated in this form. To ease the economic interpretation of our expressions some simplified notation will be introduced. The fair premium problem (13) for a separable guaranteed amount consists basically of three terms that are differently related to the distributional assumptions discussed in the previous section.[6]

- The first term denotes the cost of the contract—that is, the market value of the sum of the periodic premiums under the death distribution, which depends only on the death distribution and can be expressed by $K \cdot H$, where H is defined as

$$H := \sum_{i=0}^{n-1} D(t_0, t_i)\left[1 - \int_{t_0}^{t_i}\pi_x(t)dt\right]. \tag{17}$$

- The expected benefit of the contract can be separated in two terms. First is the expected discounted guaranteed amount under the death distribution, which depends on the specific functional form of the guaranteed amount. Due to the separability assumption (16) we can reformulate this as the function $a \cdot K \cdot L(h(K)/a \cdot K)$ with

$$L(y) := y\left\{ \int_{t_0}^{T} D(t_0, t)F(t)\pi_x(t)dt + D(t_0, T)F(T)\left[1 - \int_{t_0}^{T} \pi_x(t)dt \right]\right\}. \quad (18)$$

The second part is the expected discounted bonus under the death distribution. This term is directly related to the reference portfolio and therefore depends on both types of the financial risks. Under the separability assumption (16) we can reformulate this benefit term as the function $a \cdot K \cdot U\ (h(K)/a \cdot K)$, with

$$U(y) := \int_{t_0}^{T} c(t_0, t, y)\pi_x(t)dt + c(t_0, T, y)\left[1 - \int_{t_0}^{T} \pi_x(t)dt \right], \quad (19)$$

$$c(t_0, t, y) := D(t_0, t)E_{P^t}\left[\left[\sum_{i=0}^{n^*(t)-1} \frac{S(t)}{S(t_i)} - F(t) \cdot y \right]^+ \right] \quad \forall t > t_0.$$

The expected benefit of the contract under the death distribution is therefore determined by the sum of the functions defined by (18) and (19) and can be written as $a \cdot K \cdot R(h(K)/a \cdot K)$, with

$$R(y) := L(y) + U(y). \quad (20)$$

With this notation and the separability of the guaranteed amount, the first formulation of fair premium problem (13) is equivalent to

$$f\left(\frac{h(K^*)}{a \cdot K^*} \right) = 0, \quad (21)$$

if we assume that $a \neq 0$ and $K \neq 0$, where the function $f(\cdot)$ is defined by

$$f(y) := \frac{H}{a} - R(y) = \frac{H}{a} - L(y) - U(y). \quad (22)$$

Observe that H is independent of K, a and g and that the function $c(t_0, t, \cdot)$ fulfills the following conditions

1. $c(t_0, t, y)$ is continuous and strictly decreasing in y.
2. From equation (15) we know that $c(t_0, t, 0) = \Sigma_{i=0}^{n^*(t)-1} D(t_0, t_i)$.
3. $\lim_{y \to +\infty} c(t_0, t, y) = 0$.
4. No arbitrage and the absolute continuity of the t-forward risk-adjusted measure P^t with respect to the Lebesgue measure $\forall t$ imply that

   $$c(t_0, t, y_1) - c(t_0, t, y_2) < (y_2 - y_1)F(t)D(t_0, t) \text{ for } y_1 < y_2,$$

   from which we obtain that $U(y_1) - U(y_2) < L(y_2) - L(y_1)$ for $y_1 < y_2$.

Theorem 2. *Suppose that the death distribution is determined by the density function* $\pi_x(t)$ *and that the stochastic processes* $\{S(t)\}_{t\in[0,T]}$ *and* $\{D(t, t')\}_{t\in[0,t']}\forall t' \in [0, T]$ *are satisfying Assumptions 1 and 2.*

There exists a unique fair premium K^* *for* $a \in]0, 1[$ *if the guaranteed amount* $g(t, K)$ *can be decomposed such that*

$$g(t, K) = F(t) \cdot h(K),$$

where

i. $F(t)$ *is a nonnegative continuous and bounded function on* $[0, T]$
ii. $\tilde{h} = h(K)/K$ *is a continuous and bijective function from* $\mathbb{R}^+$ *to* $\mathbb{R}^+$.

Proof: Under the assumption made on $\tilde{h}(\cdot)$ it is sufficient to prove that there exists a unique y^* with $f(y^*) = 0$. The function $f(\cdot)$ fulfills

1. $\lim_{y\to+\infty} f(y) = -\infty$
2. for $y_1 < y_2$:

$$f(y_2) - f(y_1) = -(L(y_2) - L(y_1)) + U(y_1) - U(y_2) < 0,$$

   where the last inequality is obtained through the application of property 4 for the function $c(t_0, t, \cdot)$. The inequality implies that $f(y)$ is strictly decreasing in y.
3. By equation (15) we have that $f(0) = (1/a - 1) \cdot H > 0 \quad \forall a \in]0, 1[$.

Thus, $f(y)$ is a continuous and strictly decreasing function in y, and by the mean value theorem there exists a unique $y^* > 0$ with $f(y^*) = 0$. □

No arbitrage opportunities and a setting of equilibrium make it reasonable to introduce the condition on the g function that $g(t, K) = F(t) \cdot h(K) = F(t) \cdot \beta \cdot K$ where β is a positive constant and $F(t)$ is a continuous and bounded function on $[0, T]$. It means that the guaranteed amount is assumed to be homogeneous of degree one in the periodic premium. Typical examples of this form are

$$g(t, K) = \exp\{-\delta t\}\beta \cdot K \quad \text{or} \quad g(t, K) = \sum_{i=0}^{n^*(t)-1} \exp\{\delta(t - t_i)\}\beta \cdot K, \tag{23}$$

where $\delta > 0$ is an internal discounting rate or, respectively, growth rate guaranteed by the insurer and $\beta > 0$ is a constant. If now K^* is a fair premium and the function $g(t, K)$ is homogeneous of degree one in K, then $2K^*$ is the fair premium for the situation where the guaranteed amount is multiplied by two. At a first glance Theorem 2 cannot be applied to this situation, but a reinterpretation of the theorem immediately yields the following:

Corollary 3. *Under the assumptions of Theorem 2 and for a continuous and bounded function $F(t)$: $[0, T] \rightarrow \mathbb{R}^+$ there exists for each $a \in]0, 1[$ a unique nontrivial $\beta^* > 0$, the shape of the guaranteed amount, such that any nonnegative premium K is a solution of the fair premium problem* (21) *with the guaranteed amount*

$$g(t, K) = F(t) \cdot \beta^* \cdot K.$$

Proof: A special case of Theorem 2 is the choice of the guaranteed amount $g(t, K) = F(t) \cdot \bar{g}$ for some fixed $\bar{g} > 0$. The fair premium is defined as the unique K^* that solve the equation $f(\bar{g}/aK^*) = 0$. Denoting $\beta^* = \bar{g}/K^* > 0$ we have the wanted result. □

The corollary implies that given the function $F(t)$, there exists for each death distribution (age) a unique function for the guaranteed amount that is homogeneous of degree one in the periodic premium. For these functions of the guaranteed amount the coefficient β^* has to be estimated. Given the value of β^* the fair premium of an equity-linked life insurance contract with a guaranteed amount $F(t) \cdot \bar{g}$ is just $K^* = \bar{g}/\beta^*$, where of course the value of β^* depends on all parameters of the life insurance contract and the distributional assumptions. We now can conclude the following qualitative result with respect to the shape β of the guaranteed amount:

Proposition 4. *Let the guaranteed amount of an equity-linked life insurance contract be equal to $g(t, K) = F(t) \cdot \beta \cdot K$ where $F : [0, T] \rightarrow \mathbb{R}^+$ is a continuous and bounded function. Suppose that the solution of the fair premium problem* (21), *$\beta^*(a)$ as a function of the share $a \in]0, 1[$, is differentiable. Then the ratio $\beta^*(a)/a$ is a decreasing function in a.*

Proof: See the appendix.

The basic problem in estimating β^* from equation (21) is that we first have to approximate the options

$$E_{P^t}\left[\left[a \sum_{i=0}^{n^*(t)-1} \frac{S(t)}{S(t_i)} - F(t)\beta\right]^+\right]$$

for all $t \in [0, T]$ as a function of β and then second have to solve a zero-point problem. As the above options can be interpreted as modified Asian options, we know that there exist no closed form solutions even if we would restrict the analysis to the situation under deterministic volatility functions. In Nielsen and Sandmann [1995] we have considered the pricing of these options by modifying two approximation methods for Asian options suggested by Turnbull and Wakeman [1991] and Vorst [1992]. Comparing these approximations with a Monte Carlo simulation we reached the conclusion that neither of the methods are appropriate to estimate the coefficient β or, respectively, to estimate the fair premium K^*. On the other hand the Monte Carlo simulation technique implies that we not only have to simulate all option values but in addition have to solve the zero-point problem with respect to all simulated paths of the sums $\Sigma_{i=0}^{n^*(t)-1} S(t)/S(t_i)$. This implies a serious computational problem. To avoid this problem the zero-point problem (21) can be turned into the following dual problem:

Theorem 5. *Consider the function*

$$a : \mathbb{R}^+ \to [0, 1] \text{ with } a(b) := \frac{H}{R(b)}.$$

Then $\beta^* = a(b) \cdot b$ is the unique solution for the fair premium problem (21) with guaranteed amount $g(t, K) = F(t) \cdot K$, and the share $a(b)$ invested in the fund at each premium payment date. Furthermore $a(b)$ is decreasing and $1/a(b)$ is convex in b.

Proof: Note first that for $a(b)$ as defined above, $\beta^* = a(b) \cdot b$ is a solution to equation (21). Second, by equation (15) we know that $a(0) = 1$ and $\lim_{b \to +\infty} a(b) = 0$. Furthermore, consider the denominator of the function $a(b)$. Differentiating with respect to b yields as in the proof of Proposition 4 that the denominator is strictly increasing. Differentiating a second time gives that the denominator is also a convex function. Thus $a(b) \in]0, 1[\ \forall\, b \in \mathbb{R}^+$. By Corollary 3 we furthermore know that the solution $\beta^* = a(b)$ is unique. □

Theorem 5 expresses that if $\beta^* = a \cdot b$ is a solution of the problem (21) for some given $b \in \mathbb{R}^+$, then a must be equal to the value defined by $a(b)$. On the reverse for $a = a(b)$ we know that the β^* that solves equation (21) must be equal to $a(b) \cdot b$. Thus we can interpret $a(b)$ as the fair share. The advantage of this formulation is that the function $a(b)$ is much easier to calculate than the solution of the zero-point problem (21). Furthermore, the limits of the function $a(b)$ are quite intuitive since

- $a(0) = 1$ implies that $\beta^* = a(0) \cdot 0 = 0$. This corresponds to a contract situation with no guaranteed amount. As we already know, the fair premium in this case is a K^* that satisfies $0 = (1 - a) \cdot K \cdot H$. For $a = 1$, this is true $\forall K$.
- $\lim_{b \to +\infty} a(b) = 0$ implies that we consider an equity-linked life insurance contract with no investment in the portfolio. In this situation we know that the β^* satisfying equation (21) is just equal to

$$\beta^* = \frac{H}{\int_{t_0}^{T} D(t_0, t) F(t) \pi_x(t) dt + F(T) D(t_0, T) \left[1 - \int_{t_0}^{T} \pi_x(t) dt \right]},$$

which is by l'Hopital's rule just equal to the limit $\lim_{b \to \infty} a(b) \cdot b$.

Finally we get the following theorem:

Theorem 6. *Suppose that the guaranteed amount of an equity-linked life insurance contract is equal to $g(t, K) = F(t) \cdot \bar{g}$ for some fixed $\bar{g} > 0$ and that $F : [0, T] \to \mathbb{R}^+$ is a continuous and bounded function. Under the distributional assumptions of Theorem 2 the fair premium K^* as a function of the share a is strictly increasing and convex.*

Consequently for a fixed periodic premium K and a guaranteed amount $g(t, K) = F(t) \cdot \beta^* \cdot K$ the shape β^*, as a function of the share a, is decreasing and concave.

Proof: See the appendix.

5. Approximation techniques in a standard Gaussian framework

We now address the numerical question of calculating the fair premium. As we have pointed out in Section 4 the fair premium K^* with separated guaranteed amount is obtained by calculating the function of the fair share $a(b) = H/R(b)$.

The numerical procedure that we are going to apply essentially consists of two steps. First, we calculate the values of $a(b_i)$ for a grid of values $b_i \in \{b_1 < b_2 < \cdots\}$, and second, we approximate by interpolation the function $a(b)$. More precisely the procedure is as follows. Suppose that for a fixed $\bar{a} \in]0, 1[$ we have $b_i < b_{i+1}$ such that

$$a_{i+1} := a(b_{i+1}) < \bar{a} < a(b_i) =: a_i.$$

The function $a(b)$ we approximate on this interval by the following decreasing function:

$$\hat{x}(\bar{b}) := \frac{a_i a_{i+1}(b_{i+1} - b_i)^2}{(\sqrt{a_{i+1}}\,[b_{i+1} - \bar{b}] + \sqrt{a_i}\,[\bar{b} - b_i])^2} \quad \text{for } \bar{b} \in [b_i, b_{i+1}]. \tag{24}$$

The inverse function[7]

$$\hat{x}^{-1}(\bar{a}) = \frac{\sqrt{a_i a_{i+1}}\,[b_{i+1} - b_i] - \sqrt{\bar{a}}\,[\sqrt{a_{i+1}}\,b_{i+1} - \sqrt{a_i}\,b_i]}{\sqrt{\bar{a}}\,[\sqrt{a_i} - \sqrt{a_{i+1}}]} \quad \text{for } \bar{a} \in [a_{i+1}, a_i] \tag{25}$$

determines for a given share $\bar{a}$ the approximation of $\bar{b}$. The fair premium of an equity-linked life policy with a guaranteed amount $g(t, K) = F(t) \cdot \bar{g} \geq 0$ and the share equal to the interpolated value $\bar{a}$ is then approximated by

$$\hat{K}^*(\bar{a}) = \frac{\bar{g}}{\bar{a} \cdot \bar{b}}. \tag{26}$$

Given this two-step procedure it remains to calculate the function $a(b)$ at sufficiently many points b_i, which implies the calculation of the function $R(\cdot)$ as defined by (20) at these points. As $R(\cdot)$ has many similarities in common with an Asian option we will as an initial step apply the approximation techniques known for these options to calculate the function $R(\cdot)$. Up to now we have not used any specific distributional assumption—that is, the present results are valid for any specification of the dynamics for the mutual fund and the term structure of interest rates as long as the regularity conditions (Assumptions 1 and 2) are satisfied. Since all known methods to approximate the arbitrage price of an Asian option are restricted to the case of a deterministic and moreover constant volatility structure, we will from now on restrict ourselves to the simplest situation—that is,

$$\begin{aligned} \sigma_1(t) &= \sigma_1 \ \forall t \\ \sigma_2(t) &= \sigma_2 > 0 \ \forall t \\ \sigma(t, t') &= \sigma \cdot (t' - t) \ \forall t \in [0, t'], \ \forall t' \in [0, T], \ \sigma > 0. \end{aligned} \tag{27}$$

As a consequence the fund follows a standard geometric Brownian motion, and the term structure model is specified as the continuous time limit of the Ho and Lee [1986] model.

In Nielsen and Sandmann [1995] we have considered three different methods concerned with this problem: the fast Fourier transformation (FFT), the Turnbull and Wakeman [1991] method, and the Vorst [1992] approximation technique. We have shown that all of the three mentioned methods are for different reasons not suitable for the situation we are considering here. First, due to the assumption of stochastic interest rates, the FFT technique cannot be applied. Second, the Turnbull and Wakeman [1991] approximation leads to a completely unusable approximation of the density of the arithmetic average

$$A(t) = \frac{1}{n^*(t)} \sum_{i=0}^{n^*(t)-1} \frac{S(t)}{S(t_i)}, \tag{28}$$

given that the time to maturity of the insurance contract exceeds five years. Although we have generalized the Turnbull and Wakeman approximation to the case of stochastic interest rates,[8] we must strongly reject this approximation in the case of an equity-linked life insurance policy (see also Figures 2 and 3). Third, the Vorst [1992] approximation turns out to be a little bit more robust, but compared with the results from a Monte Carlo simulation in Section 6 this approximation is not suitable for the determination of the fair premium. To clarify this we consider the Vorst approximation in more detail.

Define in addition to the arithmetic average the geometric average by

$$G(t) = \sqrt[n^*(t)]{\prod_{i=0}^{n^*(t)-1} \frac{S(t)}{S(t_i)}} \leq A(t). \tag{29}$$

The relationship between these averages implies

$$\begin{aligned} D(t_0, t)E_{P^t}[[G(t) - Y]^+] &\leq D(t_0, t)E_{P^t}[[A(t) - Y]^+] \\ &\leq D(t_0, t)E_{P^t}[[G(t) - Y]^+] + D(t_0, t)(E_{P^t}[A(t)] - E_{P^t}[G(t)]). \end{aligned} \tag{30}$$

Under the assumption of a log-normally distributed mutual fund and a Gaussian term structure model the geometric average $G(t)$ is log-normally distributed under the t-forward risk-adjusted measure P^t. This implies that the t-forward value of the option on the geometric average is determined by

$$E_{P^t}[\max\{G(t) - Y, 0\}] = \exp\{m_G(t) + \frac{1}{2}\sigma_G^2(t)\}\, N(d) - YN(d - \sigma_G(t)) \quad (31)$$

$$d = \frac{m_G(t) - \ln Y + \frac{1}{2}\sigma_G^2(t)}{\sigma_G(t)},$$

$$m_G(t) = E_{P^t}[\ln G(t)] \quad \text{and} \quad \sigma_G^2(t) = V_{P^t}[\ln G(t)],$$

where $N(\cdot)$ denotes the standard normal distribution. In the case of a Gaussian term structure model Vorst [1992] provides us with upper and lower bounds for the Asian options that do imply lower and upper bounds for the fair premium. Furthermore, Vorst suggests the following approximation for the t-forward value of an Asian option:

$$E_{P^t}[[A(t) - Y]^+] \approx \exp\{m_G(t) + \frac{1}{2}\sigma_G^2(t)\}\, N(z) - \tilde{Y} \cdot N(z - \sigma_G(t)) \quad (32)$$

with

$$\tilde{Y} = Y - (E_{P^t}[A(t)] - E_{P^t}[G(t)]) \quad \text{and} \quad z = \frac{m_G(t) - \ln \tilde{Y} + \frac{1}{2}\sigma_G^2(t)}{\sigma_G(t)},$$

The Vorst approximation implies that the value of an Asian option is approximated by the value of an option based on the geometric average, where the log-normal distribution of the geometric average has a support $[E_{P^t}[A(t)] - E_{P^t}[G(t)], +\infty[$. Remember that the unknown distribution of the arithmetic average has the support $\mathbb{R}^+$. Therefore, the difference between the expected arithmetic and geometric average under the t-forward measure determines the approximation error. From equation (15) we already know that $E_{P^t}[A(t)] = 1/n^*(t) \sum_{i=0}^{n^*(t)} D(t_0, t_i)/D(t_0, t)$, which for a flat initial term structure with $D(t_0, t_i) = \exp\{-rt_i\}$, $r > 0$ is strictly increasing. In this situation we can calculate the expected value of the geometric average by the following:

Theorem 7. *If the stochastic processes* $\{S(t)\}_{t\in[0,T]}$ *and* $\{D(t, t')\}_{t\in[0,t']}$ $\forall t' \in [0, T]$ *are solutions of the stochastic differential equations* (8) *and, respectively,* (9) *with constant volatility coefficients, then the expected value of the geometric average* $G(t)$ *under the t-forward risk adjusted measure is equal to*

$$E_{P^t}[G(t)] = \exp\left\{ m_G(t) + \frac{1}{2}\sigma_G^2(t) \right\}$$

with

$$m_G(t) = \frac{1}{n^*(t)} \sum_{i=0}^{n^*(t)-1} \left[\ln\left(\frac{D(t_0, t_i)}{D(t_0, t)} \right) - \frac{1}{2}\left((\sigma_1^2 + \sigma_2^2)(t - t_i) + (\sigma^2 t_i - \sigma_1\sigma)(t - t_i)^2 + \frac{1}{3}\sigma^2(t - t_i)^3 \right) \right],$$

$$\sigma_G^2(t) = \frac{\sigma_1^2 + \sigma_2^2}{n^*(t)^2} \sum_{i=0}^{n^*(t)-1} (i+1)^2(t_{i+1} - t_i) - \frac{\sigma_1 \sigma}{n^*(t)^2} \sum_{i=0}^{n^*(t)-1} (i+1)^2(t_{i+1} - t_i)[2t - t_{i+1} - t_i]$$

$$+ \frac{2\sigma_1\sigma}{n^*(t)^2} \sum_{i=0}^{n^*(t)-2} \left[\left(\sum_{j=i+1}^{n^*(t)-1} (j - i)(t_{j+1} - t_j) \right) (i+1)(t_{i+1} - t_i) \right]$$

$$+ \frac{\sigma^2}{3n^*(t)^2} \sum_{i=0}^{n^*(t)-1} (i+1)^2((t - t_i)^3 - (t - t_{i+1})^3)$$

$$+ \frac{\sigma^2}{n^*(t)^2} \sum_{i=0}^{n^*(t)-2} \left[\left(\sum_{j=i+1}^{n^*(t)-1} (j - i)(t_{j+1} - t_j) \right)^2 (t_{i+1} - t_i) \right]$$

$$- \frac{\sigma^2}{n^*(t)^2} \sum_{i=0}^{n^*(t)-2} \left[\left(\sum_{j=i+1}^{n^*(t)-1} (j - i)(t_{j+1} - t_j) \right) (i+1)(t_{i+1} - t_i)[2t - t_{i+1} - t_i] \right]$$

For a constant discretization of the time axis—that is, $\Delta t = t_{i+1} - t_i$, $t = t_{n^*(t)-1} + \Delta t$ *and a flat initial term structure—this can be simplified to*

$$E_{P'}[G(t)] = \exp\left\{ \frac{1}{2} rt \frac{n(n+1)}{n^2} \right\} \cdot \exp\left\{ - \frac{\sigma_1^2 + \sigma_2^2}{12} t \frac{n^2 - 1}{n^2} \right\}$$

$$\cdot \exp\left\{ \frac{\sigma_1\sigma}{24} t^2 \frac{(n-1)(3n^2 + 7n + 4)}{n^2} \right\}$$

$$\cdot \exp\left\{ - \frac{\sigma^2}{720} t^3 \frac{48n^4 + 75n^3 - 10n^2 - 75n - 38}{n^4} \right\}.$$

Proof: See the appendix.

Remarks:

- Consider a deterministic term structure—that is, $\sigma = 0$. In this situation the expected value $E_{P'}[G(t)]$ is strictly decreasing in the volatility $\sigma_S = \sqrt{\sigma_1^2 + \sigma_2^2}$ of the underlying mutual fund. If, furthermore, we assume a flat initial term structure and an equidistant time grid, we have

$$\lim_{\Delta t \to 0} E_{P'}[G(t)] = \exp\left\{ \frac{1}{12} (6r - \sigma_S^2)t \right\},$$

  which is either decreasing (if $r < 1/6\, \sigma_S^2$, which is usually not the case) or increasing (if $r > 1/6\, \sigma_S^2$) in the maturity t of the insurance contract.

- If the interest rate is stochastic—that is, $\sigma > 0$—the properties of the expected geometric average are different. Consider the situation of a flat initial term structure and an equidistant time grid Δt. For a fixed volatility $\sigma_S = \sqrt{\sigma_1^2 + \sigma_2^2}$ of the underlying mutual fund the expected geometric average is strictly increasing in the correlation $\rho = \sigma_1/\sigma_S$. In addition for $t > 1$ the expected geometric average is a strictly decreasing function of the interest-rate volatility $\sigma > 0$—that is, a volatile interest-rate market reduces the expected geometric average. Furthermore, the expected geometric average converge toward zero in the maturity t of the insurance contract if $\sigma > 0$ and $n^*(t) > 1$ independent of the initial interest rate r.
- The expected arithmetic average is independent of the volatility functions $\sigma_1(\cdot)$, $\sigma_2(\cdot)$ and $\sigma(\cdot)$, whereas the expected geometric average is strongly dependant on these functions. For a flat initial term structure we know that

$$E_{P^t}[A(t)] = \frac{1}{n^*(t)} \sum_{i=0}^{n^*(t)-1} \exp\{r(t - t_i)\},$$

which for an equidistant time grid yields

$$\lim_{\Delta t \to 0} E_{P^t}[A(t)] = \frac{1}{t \cdot r}[e^{rt} - 1] \to +\infty \quad \text{for } t \to +\infty.$$

Thus the expected arithmetic average increases, whereas the expected geometric average approach zero for time to maturity t going to infinity. In addition we can conclude from Theorem 7 that

$$\lim_{\Delta t \to 0} V_{P^t}[G(t)] = \left(\exp\left\{-\frac{48}{120}\sigma^2 t^4\right\}\right)^2 \left(\exp\left\{-\frac{12}{360}\sigma^2 t^4\right\} - 1\right)$$

$$\to 0 \text{ for } t \to \infty \text{ and } \sigma > 0.$$

This implies that we expect, for an approximation method based on the geometric average, a systematic error for the pricing of the Asian option if the time to maturity is large. As a consequence we expect a systematic error in the pricing of the bonus of an equity-linked life insurance policy and the fair premium. Figure 1 summarizes some of the mentioned properties of the expected geometric and arithmetic average.

6. Numerical results

In this section we apply a Monte Carlo simulation to delineate our theoretical results in Sections 4 and 5. We will compare the results obtained by the Monte Carlo simulation with those derived by adopting the Vorst approximation. The life insurance policies that we consider are characterized by maturities of twelve, fifteen, and eighteen years. As functions of the guaranteed amount we restrict ourselves to two special types of a separable form (23)—that is,

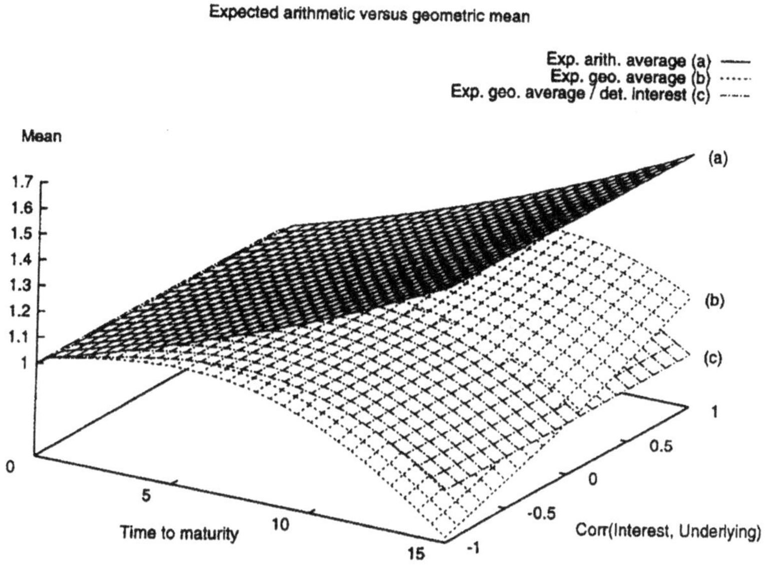

Figure 1. Expected arithmetic average $A(t)$ and geometric average $G(t)$ under the t-forward measure as a function of the maturity t and the correlation ρ for a flat initial term structure with $D(t_0, t) = \exp\{-0.0582 \cdot t\}$, $\sigma_s = 0.25$, $\sigma = 0.1$, and $\Delta t = t_{i+1} - t_i = 1/12$.

$$\text{Model 1} \quad g(t, K) = \exp\{-\delta t\}\beta \cdot K \quad \text{for} \quad \delta = 0,\ 0.02 \text{ or } 0.035, \tag{33}$$

$$\text{Model 2} \quad g(t, K) = \sum_{i=0}^{n^*(t)-1} \exp\{\delta(t - t_i)\}\beta \cdot K \quad \text{for } \delta = 0.025,\ 0.035, \text{ or } 0.045.$$

In all the simulations we use a flat initial term structure with $D(t_0, t) = \exp\{-0.0582 \cdot t\}$ and assume a monthly payment frequency for the periodic premium. The death distribution is chosen to satisfy the Makeham formula (see, e.g., Delbaen [1990])—that is,

$$\pi_x(t_i) = \frac{l_{x+t_i} - l_{x+t+\Delta t}}{l_x} \tag{34}$$

$\hat{=}$ probability that a life-aged-x will survive t_i years and die within the following Δt years,

where

$$l_x = b \cdot s^x \cdot g^{c^x},$$
$$s = 0.99949255, \quad g = 0.99959845,$$
$$c = 1.10291509, \quad b = 1000401.71,$$

and the age is fixed by $x = 30$ years. We assume that if the insured person dies at time $t \in [t_i, t_{i+1}]$ where we fixed $\Delta t = t_{i+1} - t_i = 1/12$, then the insurer pays the guaranteed amount plus the bonus at time t_{i+1}. Therefore, the estimation equation can be discretized with respect to the death distribution. As variance reduction we use an antithetic and control variate technique with the geometric average option as the control variate. The stochastic process of the underlying mutual fund and the zero coupon bonds are specified by the following simple volatility specifications:

$$\sigma_1 = 0, \sigma_2 = 25\% \text{ and } \sigma(t, t') = 0.01(t' - t).$$

That is, we concentrate on a situation where the correlation between the interest-rate market and the mutual fund is equal to zero. The calculation of the fair premium based on a Monte Carlo simulation is determined by the two step procedure introduced in Section 5. Thus we simulate the function $a(\cdot)$ defined in Theorem 5 by

$$\hat{a}(b) = \frac{H}{\hat{R}(b)} \tag{35}$$

$$\hat{R}(b) = b\left(\sum_{i=0}^{n-1} D(t_0, t_{i+1})F(t_i)\pi(t_i) + F(t_n)D(t_0, t_n)\left(1 - \sum_{i=0}^{n-1} \pi(t_i)\right)\right) \tag{36}$$

$$+ \sum_{i=0}^{n-1} \hat{c}(t_0, t_i, b)\pi(t_i) + \hat{c}(t_0, t_n, b)\left(1 - \sum_{i=0}^{n-1} \pi(t_i)\right),$$

$$\hat{c}(t_0, t, b) = \frac{D(t_0, t)}{2M} \sum_{m=1}^{2M} \left[\left[\sum_{j=0}^{n^*(t)-1} \frac{S_m(t)}{S_m(t_j)} - F(t) \cdot b\right]^+\right.$$

$$\left. - n^*(t) \cdot \left[\sqrt[n^*(t)]{\prod_{j=0}^{n^*(t)-1} \frac{S_m(t)}{S_m(t_j)}} - F(t) \cdot \frac{b}{n^*(t)}\right]^+\right]$$

$$+ n^*(t) \cdot D(t_0, t) \cdot E_{P^t}\left[\left[G(t) - F(t) \cdot \frac{b}{n^*(t)}\right]^+\right],$$

where M denotes the number of paths[9] and the value of the geometric average option is given by equation (31). We set $\Delta t = t_{i+1} - t_i = 1/12$, which implies that n is equal to 144,180, or 216, depending on the maturity of the insurance contract (twelve, fifteen, or eighteen years). Note that we have to simulate between 144 and 216 option values for the mentioned maturities. To compare the Monte Carlo results we determine the boundary solutions and the approximation for Asian options suggested by Vorst [1992]. Instead of estimating the function $a(b)$ we solve the zero-point problem (21) directly applying the closed-form solutions for the related geometric average option as discussed in Section 5.

Figures 2 and 3 show the different approximations of the p.d.f. of the arithmetic average. As already mentioned in Section 5 the Turnbull and Wakeman [1991] method leads to a completely unreasonable approximation. The Vorst [1992] approximation seems to be more valid if we consider a time to maturity equal to three years. For the case of nine years

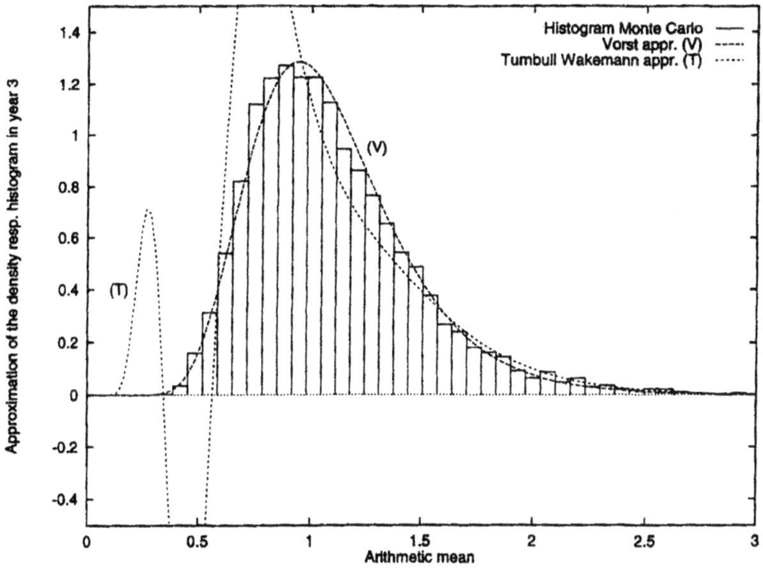

Figure 2. Approximation of the p.d.f. for the arithmetic average $A(t)$ after three years. Equity-linked life insurance contract with monthly payment frequency; flat initial term structure, age of the insured person = 30 years; $\sigma_1 = 0$; $\sigma_2 = 0.25$; $\sigma = 0.1$; Monte Carlo simulation with $M = 6000$.

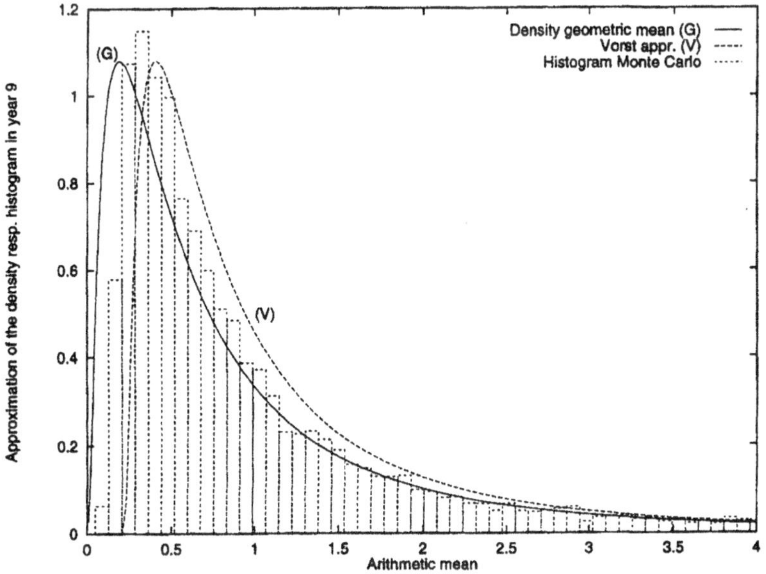

Figure 3. Approximation of the p.d.f. for the arithmetic average $A(t)$ after nine years. Equity-linked life insurance contract with monthly payment frequency; flat initial term structure, age of the insured person = 30 years; $\sigma_1 = 0$; $\sigma_2 = 0.25$; $\sigma = 0.1$; Monte Carlo simulation with $M = 6000$.

we recognize the shift of the density support measured by the difference between the expected arithmetic and geometric average. The p.d.f. for the geometric average represents the lower bound for the arithmetic average. The Vorst approximation seems to systematically overestimate the probability distribution. We therefore expect that the approximation of the fair premium based on the Vorst method will systematically underestimate the Monte Carlo result.

In Figures 4 and 5 we show the approximation of the function $a(b)$ defined in Theorem 5 for different parameter situations. For a fixed share $\bar{a} = a(b)$ the value of b is increasing in time to maturity. This implies a lower premium of the insurance contract if the time to maturity increases. Furthermore, if the discount rate δ increases, the fair premium for an otherwise equal life insurance contract decreases. A discount rate $\delta > 0$ in Model 1 implies that the guaranteed amount decreases in time, whereas $\delta = 0$ indicates a time-independent guaranteed amount.

The situation for Model 2 is different. The function of the guaranteed amount is increasing in time and in the internal growth rate δ. If the growth rate is fixed (say, 4.5 percent) as in Figure 6, the fair premium depends on two opposite effects. As the time to maturity of the insurance contract increases, the fair premium will decrease due to the expected delay in the payment of the benefit. On the other side, if the time to maturity increases, the guaranteed amount will increase, which on its own implies a higher premium. As Figure 6 indicates, these two effects add up in such a manner that the function $a(b)$ is nearly independent of the time to maturity. It means that the fair premiums for these contracts are very similar to each other. If the maturity of the contract is fixed as in Figure 7, the functions

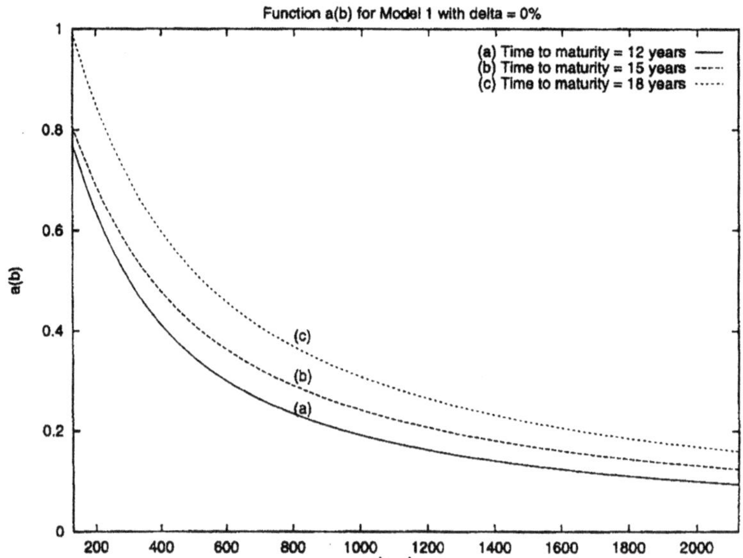

Figure 4. The function $a(b)$ in Model 1 with $\sigma = 0.0\%$ and different times to maturity for an equity-linked life insurance contract with monthly payment frequency; age of the insured = 30 years; $\sigma_1 = 0$; $\sigma_2 = 0.25$; $\sigma = 0.1$; $M = 6000$; and a flat initial term structure.

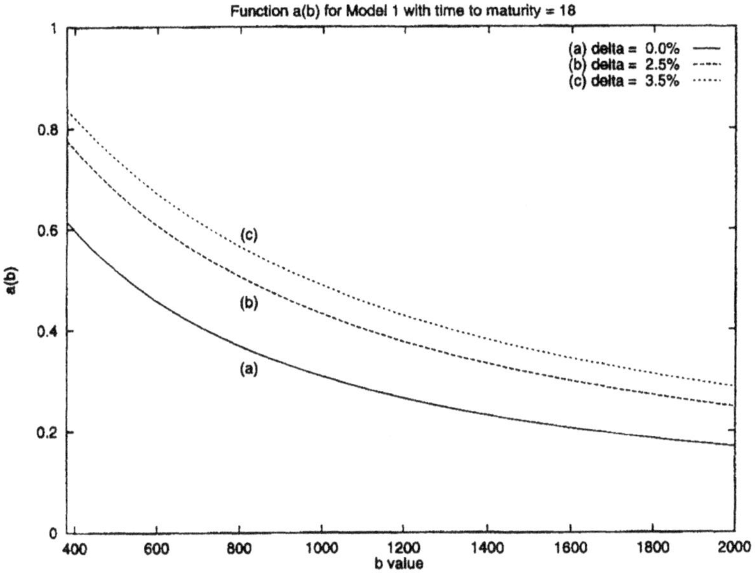

Figure 5. The function $a(b)$ in Model 1 with time to maturity eighteen years and different value for the internal interest rate δ for an equity-linked life insurance contract with monthly payment frequency; age of the insured person = 30 years; $\sigma_1 = 0$; $\sigma_2 = 0.25$; $\sigma = 0.1$; $M = 6000$; and a flat initial term structure.

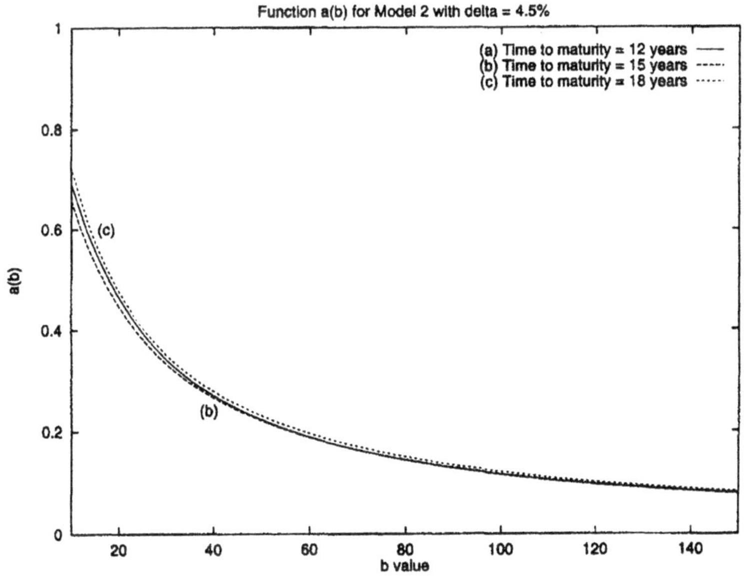

Figure 6. The function $a(b)$ in Model 2 with $\delta = 4.5\%$ and different times to maturity for an equity-linked life insurance contract with monthly payment frequency; age of the insured person = 30 years; $\sigma_1 = 0$; $\sigma_2 = 0.25$; $\sigma = 0.1$; $M = 6000$; and a flat initial term structure.

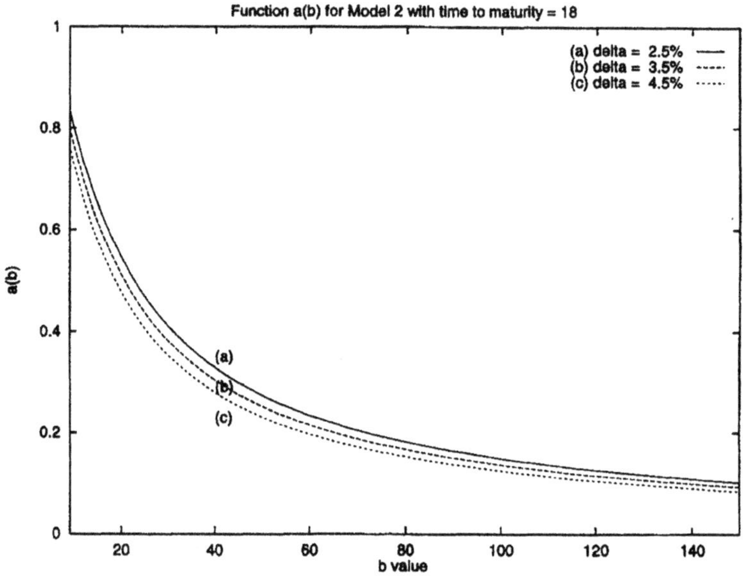

Figure 7. The function $a(b)$ in Model 2 with time to maturity eighteen years and different value for the internal growth rate δ for an equity-linked life insurance contract with monthly payment frequency; age of the insured person $= 30$ years; $\sigma_1 = 0$; $\sigma_2 = 0.25$; $\sigma = 0.1$; $M = 6000$; and a flat initial term structure.

$a(b)$ are decreasing in the internal growth rate δ. This implies that the fair premium is increasing in the growth rate. This makes sense; if everything else remains equal, these contracts differ in a monotone manner in the size of the guaranteed amount. In addition to these direct effects of the growth rate, there is an indirect effect, since an increase in the guaranteed amount implies a decrease of the bonus, and therefore the value of the weighted options will decrease. This indirect effect will on average decrease the fair premium and may be responsible, as Figure 6 suggests, for the small difference between the simulated functions $a(b)$ across different internal growth rates.

Tables 1 and 2 show within Model 1 simulated values for the fair premium obtained by the Monte Carlo simulation. Since Model 1 implies a decreasing guaranteed amount if the internal growth rate δ exceeds zero, we can either consider a fixed guaranteed amount at maturity of the contract (Table 1) or at the beginning of the contract (Table 2). In both cases the fair premium is an increasing function of the share a, which is in accordance with Theorem 6. This is a general property within the diffusion framework considered. In addition the fair premium decreases remarkably with the discount rate δ if the guaranteed amount is fixed in the beginning. This is perfectly in line with the contract specifications in Model 1. If instead the guaranteed amount is fixed at the maturity of the contract, an internal growth rate $\delta > 0$ implies that the guaranteed amount at time t_0 increases in δ. Thus in the case of the death of the insured person the insurer guarantees a higher amount as with $\delta = 0$. This results as Table 2 shows in an increasing premium. Nevertheless we have to note that this increase is of quite small magnitude. The reason is that a premature payment is not likely under the chosen death distribution and the fair premium is mainly determined by the payment at maturity, which in Table 2 is fixed for all contracts.

Table 1. Fair premiums in Model 1 with a fixed guaranteed amount of 10.000 at time 0.

Maturity	12			18		
Delta	0.0%	2.5%	3.5%	0.0%	2.5%	3.5%
$G(T, K)$	10000.00	7408.1822	6570.4682	10000.00	6376.2815	5325.9180
$a = 0.3$	55.5872	41.3973	36.8048	32.2516	20.9222	17.6232
	(0.5219)	(0.3887)	(0.3455)	(0.7864)	(0.5101)	(0.4297)
$a = 0.35$	57.6269	42.9188	38.1585	33.4705	21.7139	18.2905
	(0.6312)	(0.4701)	(0.418)	(0.9521)	(0.6177)	(0.5203)
$a = 0.4$	59.9791	44.6741	39.7205	34.816	22.5875	19.0968
	(0.7508)	(0.5592)	(0.4972)	(1.1319)	(0.7344)	(0.6186)
$a = 0.45$	62.7108	46.7122	41.5343	36.3037	23.5546	19.8422
	(0.8831)	(0.6578)	(0.5849)	(1.3279)	(0.8615)	(0.7258)
$a = 0.5$	65.8899	49.0836	43.6442	37.9633	24.6325	20.7509
	(1.031)	(0.768)	(0.6829)	(1.5429)	(1.0011)	(0.8433)
$a = 0.55$	69.6356	51.8801	46.1333	39.8243	25.8423	21.7712
	(1.1985)	(0.8929)	(0.794)	(1.7804)	(1.1553)	(0.9733)
$a = 0.6$	74.1361	55.2377	49.1215	41.9256	27.2064	22.9207
	(1.3919)	(1.0371)	(0.9223)	(2.0447)	(1.3268)	(1.1178)
$a = 0.65$	79.5795	59.3016	52.7386	44.3083	28.7557	24.2274
	(1.6186)	(1.2062)	(1.0727)	(2.341)	(1.5193)	(1.28)
$a = 0.7$	86.3402	64.3469	57.2291	47.0475	30.5352	25.7278
	(1.8912)	(1.4094)	(1.2535)	(2.6769)	(1.7374)	(1.4639)
$a = 0.75$	94.9961	70.8121	62.9852	50.2187	32.5954	27.4648
	(2.2293)	(1.6617)	(1.4781)	(3.0615)	(1.9871)	(1.6743)
$a = 0.8$	106.6261	79.4971	70.7178	53.9374	35.9115	29.502
	(2.6689)	(1.9898)	(1.7701)	(3.5074)	(2.2767)	(1.9184)
$a = 0.85$	123.4692	92.0944	81.9388	58.366	37.8902	31.9301
	(3.2833)	(2.449)	(2.1789)	(4.0325)	(2.6179)	(2.2061)

Note: The number in parentheses is the standard deviation.

Tables 3 and 4 concentrate on the influence on the fair premium of the internal growth rate δ. As shown by Theorem 6 the fair premium is an increasing function of the share a. For a fixed guaranteed amount at time t_0 the monthly fair premium is an increasing function of the internal growth rate. Since we already know that mainly the terminal payment at maturity contributes to the fair premium, this indicates in addition that the bonus at the maturity of the contract can on average not compensate for the higher guaranteed amount. In other words on average the embedded option ends out of the money for higher internal growth rates. The value of the embedded option is therefore small and does not effect the value of the premium in any significant manner.

If the guaranteed amount is fixed at the maturity of the contract, a higher internal growth rate implies a lower guaranteed amount at the intermediate times. This should result in a lower premium, but as Table 4 shows this premium reduction is rather small. Again this is mainly based on the relative small contribution of the death distribution to the premium. In addition given the low guaranteed amount during the starting period of the contract, we can expect that the value of the reference portfolio is likely to exceed the guaranteed amount—that is, even in the case of a premature death of the insured person the payment will be quite similar for the different internal growth rates δ.

Table 2. Fair premiums in Model 1 with a fixed guaranteed amount of 10.000 at time *T.*

Maturity	12			18		
Delta	0.0%	2.5%	3.5%	0.0%	2.5%	3.5%
$G(0, K)$	10000.00	13498.5881	15219.6156	10000.00	15683.1219	18776.1058
$a = 0.3$	55.5872	55.8805	56.0155	32.2516	32.8126	33.0895
	(0.5219)	(0.5246)	(0.5259)	(0.7864)	(0.8001)	(0.8068)
$a = 0.35$	57.6269	57.9343	58.0758	33.4705	34.0542	34.3425
	(0.6312)	(0.6346)	(0.6361)	(0.9521)	(0.9687)	(0.9769)
$a = 0.4$	59.9791	60.3037	69.4531	34.816	35.4243	35.7249
	(0.7508)	(0.7549)	(0.7568)	(1.1319)	(1.1517)	(1.1615)
$a = 0.45$	62.7108	63.0549	63.2136	36.3037	36.9409	37.2559
	(0.8831)	(0.888)	(0.8902)	(1.3279)	(1.3512)	(1.3627)
$a = 0.5$	65.8899	66.2559	66.4249	37.9633	38.6315	38.9621
	(1.031)	(1.0367)	(1.0394)	(1.5429)	(1.57)	(1.5835)
$a = 0.55$	69.6356	70.0308	71.2131	39.8243	40.5289	40.8778
	(1.1985)	(1.2053)	(1.2085)	(1.7804)	(1.8119)	(1.8275)
$a = 0.6$	74.1361	74.5631	74.761	41.9256	42.6681	43.0361
	(1.3919)	(1.4)	(1.4037)	(2.0447)	(2.0809)	(2.0989)
$a = 0.65$	79.5795	80.0488	80.2662	44.3083	45.098	45.4896
	(1.6186)	(1.6282)	(1.6326)	(2.341)	(2.3827)	(2.4034)
$a = 0.7$	86.3402	86.8593	87.1005	47.0475	47.8887	48.3067
	(1.8912)	(1.9025)	(1.9078)	(2.6769)	(2.7248)	(2.7486)
$a = 0.75$	94.9961	95.5864	95.8611	50.2187	51.1198	51.5682
	(2.2293)	(2.2431)	(2.2496)	(3.0615)	(3.1164)	(3.1437)
$a = 0.8$	106.6261	107.3098	107.6297	53.9374	54.9089	55.3933
	(2.6689)	(2.686)	(2.694)	(3.5074)	(3.5705)	(3.602)
$a = 0.85$	123.4692	124.3144	124.7077	58.366	59.4237	59.9524
	(3.2833)	(3.3058)	(3.3163)	(4.0325)	(4.1056)	(4.1421)

Note: The number in parentheses is the standard deviation.

For a homogeneous guaranteed amount of the type $g(t, K) = F(t)\beta^* \cdot K$, Figures 8 and 9 represent the guaranteed amount at maturity of the contract as a function of the monthly premium for different values of the share a.

For a fixed periodic premium K the guaranteed amount is shown to be a decreasing function of the share a that results from the decrease of β^* as a function of the share a. This property of β^* is by Theorem 6 satisfied for a general diffusion framework and not restricted to the constant volatility case we have simulated. As a consequence of this, the fair premium increases as a function of the share a for a fixed guaranteed amount. Since $\beta^*(\cdot)$ as a function of the share a is concave, the same is true for the guaranteed amount.

Finally, Figures 10 and 11 show the fair premium as a function of the share a. In line with Theorem 6 the fair premium is a convex function with respect to the share a. Furthermore, the upper bound for the premium, which corresponds to the lower bounds for the options given by Vorst [1992], is very weak. This indicates that the geometric average option is not suitable to approximate the bonus part. This result is perfectly in line with our discussion of the expected geometric average in Section 5. As expected from Figure 3 the Vorst approximation of the embedded options lead to a systematic overestimation of the

Table 3. Fair premiums in Model 2 with a fixed guaranteed amount of 10.000 at time 0.

Maturity	12			18		
Delta	2.5%	4.5%	4.5%	2.5%	3.5%	4.5%
$G(T, K)$	15170.0106	16175.6904	17271.952	24017.8275	26515.9529	29359.9221
$a = 0.3$	82.9338	88.4044	94.3666	74.7204	82.409	91.1558
	(0.7787)	(0.83)	(0.886)	(1.8219)	(2.0093)	(2.2226)
$a = 0.35$	85.9634	91.634	97.8138	77.5416	85.521	94.5984
	(0.9416)	(1.0038)	(1.0714)	(2.2059)	(2.4328)	(2.6911)
$a = 0.4$	89.457	95.3572	101.789	80.6594	88.9595	98.4032
	(1.1199)	(1.1937)	(1.2742)	(2.6224)	(2.8923)	(3.1993)
$a = 0.45$	93.5132	99.6819	106.4051	84.1041	92.7606	102.6081
	(1.3169)	(1.4038)	(1.4985)	(3.0763)	(3.3929)	(3.7531)
$a = 0.5$	98.243	104.7239	111.7879	87.9511	97.0049	107.3055
	(1.5372)	(1.6387)	(1.7492)	(3.5744)	(3.9424)	(4.361)
$a = 0.55$	103.802	110.65	118.1138	92.2614	101.7604	112.5683
	(1.7866)	(1.9045)	(2.0329)	(4.1246)	(4.5493)	(5.0324)
$a = 0.6$	110.4935	117.7834	125.7292	97.1423	107.1478	118.5311
	(2.0746)	(2.2115)	(2.3606)	(4.7376)	(5.2256)	(5.7807)
$a = 0.65$	118.5912	126.4187	134.9505	102.6721	113.2477	125.2813
	(2.4121)	(2.5713)	(2.7448)	(5.4246)	(5.9834)	(6.6191)
$a = 0.7$	128.662	137.1568	146.4167	109.038	120.2741	133.06
	(2.8182)	(3.0042)	(3.207)	(6.2041)	(6.8434)	(7.5709)
$a = 0.75$	141.5566	150.9113	161.1079	116.4222	128.4242	142.0857
	(3.3219)	(3.5414)	(3.7807)	(7.0974)	(7.8291)	(8.6619)
$a = 0.8$	158.9174	169.4251	180.8776	125.0941	137.9992	152.684
	(3.9777)	(4.2407)	(4.5273)	(8.1344)	(8.9736)	(9.9285)
$a = 0.85$	184.0775	196.2225	209.5478	135.4379	149.4225	165.3376
	(4.8949)	(5.2178)	(5.5723)	(9.3575)	(10.3237)	(11.4233)

Note: The number in parentheses is the standard deviation.

embedded options and thus to a systematic underestimation of the fair premium. The size of this approximation error increases with the invested share a, since the value of the reference portfolio $a \cdot K\Sigma_{i=0}^{n^*(t)-1} S(t)/S(t_i)$ increases in a and therefore the option value will increase.

7. Conclusions

An equity-linked life insurance contract combines the technical risk of a life insurance with the financial risk of an investment strategy. More precisely, the contract offers a life insurance policy in combination with an investment strategy together with a minimum guaranteed value. From this perspective the equity-linked life insurance contract represents an interesting contract that could have a promising market perspective. Since the investment strategy is based on a mutual fund, the insured person participates in the economic growth covered by the fund, whereas the downside risk is covered by the minimum guaranteed value. On the other hand this contract specification implies a complex payoff structure.

Table 4. Fair premiums in Model 2 with a fixed guaranteed amount of 10.000 at time T.

Maturity	12			18		
Delta	2.5%	4.5%	4.5%	2.5%	3.5%	4.5%
$G(0, K)$	659.1963	618.2116	578.9734	416.3574	377.1315	340.6004
$a = 0.3$	54.6696	54.6527	54.6357	31.1104	31.079	31.0477
	(0.5133)	(0.5131)	(0.513)	(0.7586)	(0.7578)	(0.757)
$a = 0.35$	56.6667	56.6492	56.6316	32.285	32.2527	32.2202
	(0.6207)	(0.6205)	(0.6203)	(0.9184)	(0.9175)	(0.9166)
$a = 0.4$	58.9697	58.9509	58.9331	33.5831	33.5494	33.5162
	(0.7382)	(0.738)	(0.7378)	(1.0919)	(1.0908)	(1.0897)
$a = 0.45$	61.6435	61.6245	71.6057	35.0174	34.983	34.9484
	(0.8681)	(0.8679)	(0.8676)	(1.2808)	(1.2796)	(1.2783)
$a = 0.5$	64.7613	64.7415	64.7222	36.6191	36.5836	36.5483
	(1.0133)	(1.013)	(1.0127)	(1.4882)	(1.4868)	(1.4854)
$a = 0.55$	68.4258	68.4051	68.3847	38.4137	38.3771	38.3408
	(1.1777)	(1.1774)	(1.177)	(1.7173)	(1.7157)	(1.714)
$a = 0.6$	72.8368	72.8151	72.7938	40.4559	40.4088	40.3717
	(1.3676)	(1.3672)	(1.3668)	(1.9725)	(1.9707)	(1.9689)
$a = 0.65$	78.1747	78.1535	78.1328	42.7483	42.7093	42.6708
	(1.59)	(1.5896)	(1.5892)	(2.2586)	(2.2565)	(2.2545)
$a = 0.7$	84.8134	84.792	84.7714	45.3988	45.3592	45.3203
	(1.8577)	(1.8572)	(1.8568)	(2.5831)	(2.5809)	(2.5787)
$a = 0.75$	93.3135	93.2951	93.2772	48.4732	48.4328	48.3944
	(2.1898)	(2.1893)	(2.1889)	(2.9551)	(2.9526)	(2.9503)
$a = 0.8$	104.7576	104.7496	104.7233	52.0838	52.0438	52.0042
	(2.6221)	(2.6216)	(2.6212)	(3.3868)	(3.3842)	(3.3817)
$a = 0.85$	121.343	121.3071	121.3226	56.3906	56.3519	56.314
	(3.2267)	(3.2257)	(3.2262)	(3.8961)	(3.8934)	(3.8908)

Note: The number in parentheses is the standard deviation.

Due to the usually long time to maturity of life insurance policies the analysis of this contract has to take into consideration three different sources of uncertainty. Beside the technical risk specified by the death distribution of the insured, two different financial risks are effecting the value of the insurance premium: the price risk of the underlying mutual fund and the interest-rate risk. In the context of a general diffusion model for these financial risks the rational interrelationship between the periodic premium, the share of the premium invested in the mutual fund and the function specifying the guaranteed amount is analyzed. As a main result it was found that the fair periodic premium is an increasing and convex function of the share invested in the mutual fund. This result is valid even if we allow for state-dependent volatility processes.

The definition of the fair premium is based on the independence between the death distribution and the financial processes and, more important, on the ability of the insurer to perfectly diversify the technical insurance risk. Given the at present relatively small markets for this contract, this assumption seems to be questionable. One way to take care of this would be to introduce loading factors on the death distribution.

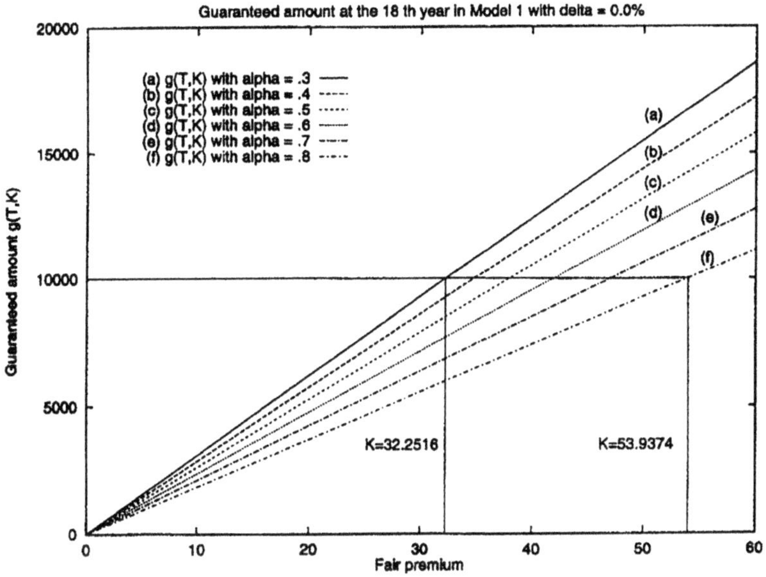

Figure 8. The guaranteed amount in Model 1 at the maturity of eighteen years as a function of the periodic premium. Equity-linked life insurance contract with monthly payment; age of the insured person = 30 years; $\sigma_1 = 0$; $\sigma_2 = 0.25$; $\sigma = 0.1$; $M = 6000$; and a flat initial term structure.

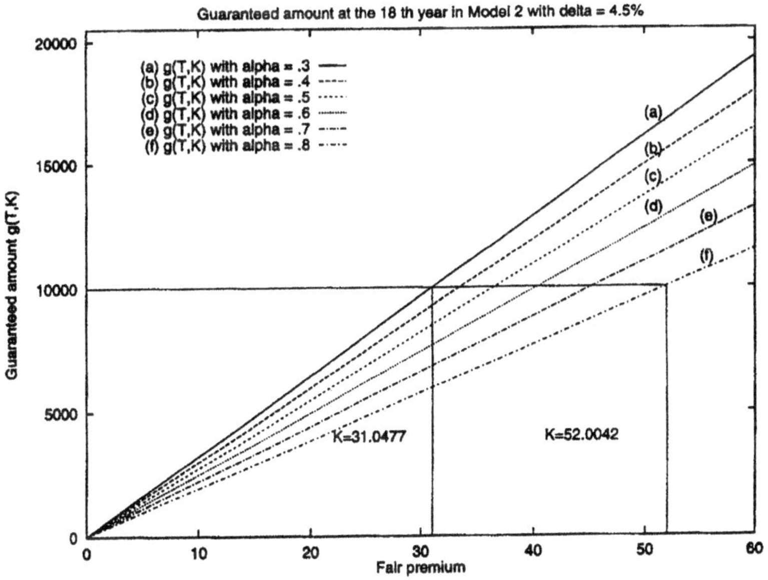

Figure 9. The guaranteed amount in Model 2 at the maturity of eighteen years as a function of the periodic premium. Equity-linked life insurance contract with monthly payment; age of the insured person = 30 years; $\sigma_1 = 0$; $\sigma_2 = 0.25$; $\sigma = 0.1$; $M = 6000$; and a flat initial term structure.

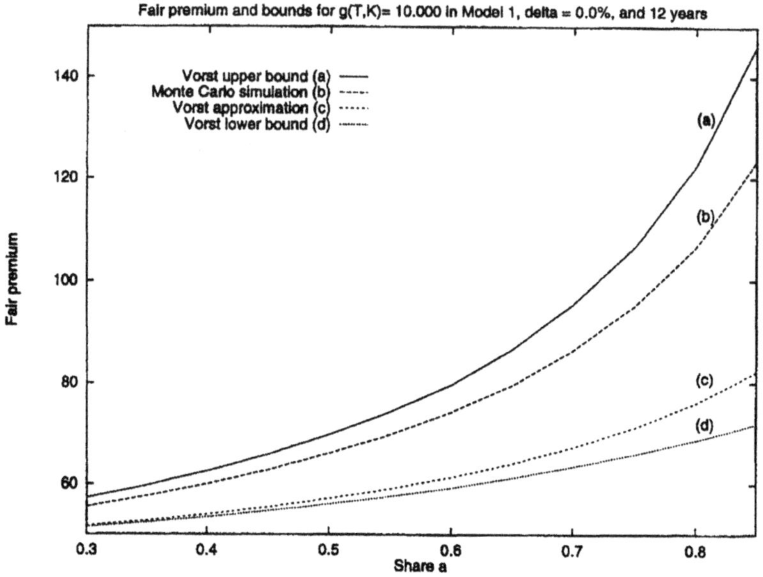

Figure 10. The fair premium and bounds implied by the Vorst approximation in Model 1 with a fixed guaranteed amount of 10.000 at the maturity of twelve years. Equity-linked life insurance contract with monthly payment frequency; age of the insured person = 30 years; $\sigma_1 = 0$; $\sigma_2 = 0.25$; $\sigma = 0.1$; $M = 6000$; and a flat initial term structure.

The analysis of this study concentrated on the properties and calculation of the fair periodic premium of an equity-linked life insurance contract. The mentioned results are obtained by the application of the arbitrage pricing theory. As an important tool we have used the change of measure technique and thus calculated the fair periodic premium under the forward risk-adjusted measures. Despite the theoretical elegance, this approach is first of all concentrated on the pricing of financial assets. It therefore does not directly address the hedging question, which is of principal importance. Due to the complexity of the contract there is no straightforward answer to this problem. Like the fair premium, the hedging problem also is related to the technical insurance risk and the financial risks. Even if we concentrate only on the financial risks the similarity betwen the insurance situation and the pricing of Asian options implies a number of problems. First is the inappropriateness for the insurance situation of those methods traditionally used for the pricing of Asian options. The failure of these methods happens because of the nondeterministic bond market and because of the long time to maturity of the insurance contracts. Our simulation results have demonstrated this in the situation of a constant volatility structure. Second, due to the long time to maturity of the insurance contracts we should expect that the volatility processes are by no means deterministic. Even more problematic, the correlation process highly influences the contract. Third, additional features such as the possibility of changing the fund or the right to cancel the contract may strongly influence the pricing and hedging of the contract.

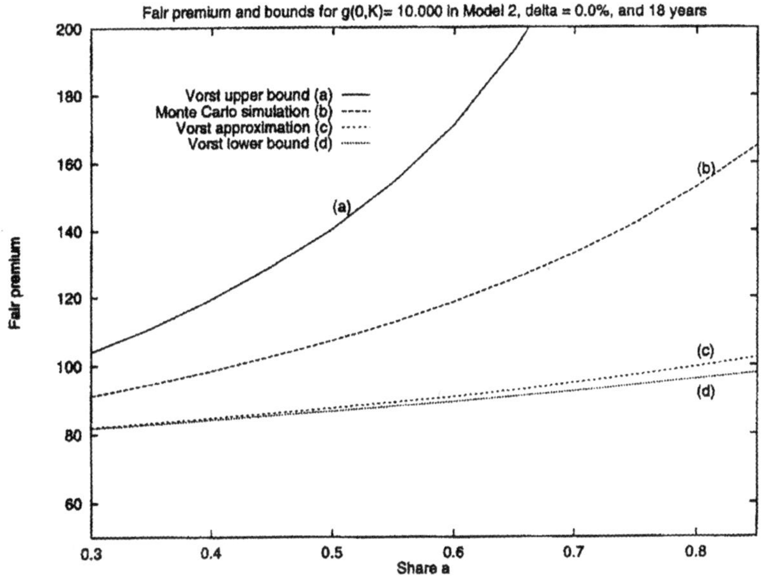

Figure 11. The fair premium and bounds implied by the Vorst approximation in Model 2 with a fixed guaranteed amount of 10.000 at the maturity of eighteen years. Equity-linked life insurance contract with monthly payment frequency; age of the insured person = 30 years; $\sigma_1 = 0$; $\sigma_2 = 0.25$; $\sigma = 0.1$; $M = 6000$; and a flat initial term structure.

We have the impression that the combination of the insurance theory and the finance theory is a fruitful and promising area for practical application and theoretical research. For an equity-linked life insurance contract with periodic premium we have discussed some relevant aspects, but perhaps unavoidably, more questions are raised than answered.

Appendix

Proof of Proposition 4

If $\beta^*(a)$ is a solution of $f(\beta^*(a)/a) = 0$ for $a \in]0, 1[$, then $\beta^*(a)$ solves

$$\frac{H}{a} = R\left(\frac{\beta^*(a)}{a}\right).$$

This implies that $\partial/\partial a(R(\beta^*(a)/a)) = -H/a^2 < 0$, and as $R(y) = L(y) + U(y)$ we have that

$$\frac{\partial R(y)}{\partial y} = \frac{\partial L(y)}{\partial y} + \frac{\partial U(y)}{\partial y}$$

$$= \frac{L(y)}{y} + \int_{t_0}^{T} \frac{\partial}{\partial y} c(t_0, t, y)\pi_x(t)dt + \frac{\partial}{\partial y} c(t_0, T, y)\left[1 - \int_{t_0}^{T} \pi_x(t)dt\right]$$

with

$$\frac{\partial}{\partial y} c(t_0, t, y) = -D(t_0, t) \int_{yF(t)}^{+\infty} \mathrm{F}(t) dQ_t(z) > -D(t_0, t)F(t) \quad \forall y > 0,$$

where $Q_t(z)$ is the unknown distribution function of the sum $\Sigma_{i=0}^{n^*(t)-1} S(t)/S(t_i)$. Therefore,

$$\frac{\partial U(y)}{\partial y} > -\int_{t_0}^{T} D(t_0, t)F(t)\pi_x(t)dt - D(t_0, T)F(T)\left[1 - \int_{t_0}^{T} \pi_x(t)dt\right] = -\frac{L(y)}{y},$$

which implies that $\partial R(y)/\partial y > 0$. Since

$$-\frac{H}{a^2} = \frac{\partial}{\partial a}\left(R\left(\frac{\beta^*(a)}{a}\right)\right) = \frac{\partial}{\partial a}\left(\frac{\beta^*(a)}{a}\right) \cdot \frac{\partial R}{\partial y}\bigg|_{y=\frac{\beta^*(a)}{a}},$$

we obtain that $\beta^*(a)/a$ is decreasing in a. □

Proof of Theorem 6

Consider the guaranteed amount $g(t, K) = F(t) \cdot \beta^* \cdot K$ where β^* is a solution of (21). Then the fair premium for the above contract is equal to $K^* = \bar{g}/\beta^*$. Let $a_1 < a_2$ and $a_1, a_2 \in]0, 1[$. Then there exists $b_1, b_2 \in \mathbb{R}$ such that the function $a(b)$ in Theorem 5 satisfies $a(b_1) = a_1$ and $a(b_2) = a_2$. Since $a(b)$ is a strictly decreasing function, $a_1 < a_2$ implies that $b_1 < b_2$. Furthermore, we can write

$$K^*(a_2) - K^*(a_1) = \frac{\bar{g}}{\beta^*(a_2)} - \frac{\bar{g}}{\beta^*(a_1)} = \bar{g}\,\frac{\beta^*(a_1) - \beta^*(a_2)}{\beta^*(a_2)\beta^*(a_1)},$$

where

$$\beta^*(a_1) - \beta^*(a_2) = a(b_1) \cdot b_1 - a(b_2) \cdot b_2.$$

We therefore have to prove that $a(b_1) \cdot b_1 > a(b_2) \cdot b_2 \; \forall b_1, b_2$—that is, we want to prove that the function $a(b) \cdot b$ is increasing in b. The condition we are looking for can then be written as

$$\frac{\partial}{\partial b}(a(b) \cdot b) = \frac{\partial}{\partial b}\left(\frac{H \cdot b}{R(b)}\right) = H \cdot \frac{R(b) - b \cdot \dfrac{\partial R(b)}{\partial b}}{R^2(b)} > 0,$$

$$\Leftrightarrow \quad R(b) - b \cdot \frac{\partial R(b)}{\partial b} > 0$$

which indeed is the case since

$$R(b) - b \cdot \frac{\partial R(b)}{\partial b} = U(b) - b \cdot \frac{\partial U(b)}{\partial b} > 0 \text{ and } \frac{\partial U(b)}{\partial b} < 0.$$

For the convexity, consider for $h > 0$ and $a \in]0, 1 - 2h]$

$$\frac{1}{h}\left(\frac{K^*(a+2h) - K^*(a+h)}{h} - \frac{K^*(a+h) - K^*(a)}{h}\right)$$

$$= \frac{\bar{g}}{h^2}\left(\frac{1}{\beta^*(a+2h)} - \frac{2}{\beta^*(a+h)} + \frac{1}{\beta^*(a)}\right)$$

$$= \frac{\bar{g}}{h^2}\left[\frac{\beta^*(a)[\beta^*(a+h) - \beta^*(a+2h)] - \beta^*(a+2h)[\beta^*(a) - \beta^*(a+h)]}{\beta^*(a)\beta^*(a+h)\beta^*(a+2h)}\right]$$

$$> \frac{\bar{g}}{h^2}\beta^*(a)\left[\frac{[\beta^*(a+h) - \beta^*(a+2h)] - [\beta^*(a) - \beta^*(a+h)]}{\beta^*(a)\beta^*(a+h)\beta^*(a+2h)}\right]$$

$$= \frac{-\bar{g}\beta^*(a)}{\beta^*(a)\beta^*(a+h)\beta^*(a+2h)}\left[\frac{\beta^*(a+2h) - \beta^*(a+h)}{h^2} - \frac{\beta^*(a+h) - \beta^*(a)}{h^2}\right],$$

since $\beta^*(a)$ is decreasing in a and positive. Therefore it is sufficient to prove that β^* is a concave function in a. As before this is equivalent to the concavity of the function $a(b) \cdot b$ for $b \in R_{\geq 0}$.

$$\frac{\partial^2(a(b) \cdot b)}{\partial b^2} = H \cdot \frac{R(b)\left(-b \cdot \frac{\partial^2 R(b)}{\partial b^2}\right) - \left(R(b) - b \cdot \frac{\partial R(b)}{\partial b}\right) 2\frac{\partial R(b)}{\partial b}}{R(b)^3} < 0$$

since

$$\frac{\partial R(b)}{\partial b} > 0, \frac{\partial^2 R(b)}{\partial b^2} > 0 \text{ and } \left(R(b) - b \cdot \frac{\partial R(b)}{\partial b}\right) > 0. \qquad \square$$

Proof of Theorem 7

For simplicity set $n := n^*(t)$. Due to the log-normality assumption the expected value of the logarithmic ratio $S(t)/S(t_i)$ is determined by

$$E_{P^t}\left[\ln\left(\frac{S(t)}{S(t_i)}\right)\right] = \ln\left(\frac{D(t_0, t_i)}{D(t_0, t)}\right) - \frac{1}{2}(t - t_i)^2\sigma^2 t_i - \frac{1}{2}\int_{t_i}^{t}((\sigma_1 - (t-u)\sigma)^2 + \sigma_2^2)du$$

$$= \ln\left(\frac{D(t_0, t_i)}{D(t_0, t)}\right) - \frac{1}{2}(t - t_i)^2\sigma^2 t_i - \frac{1}{2}(\sigma_1^2 + \sigma_2^2)(t - t_i)$$

$$+ \frac{1}{2}(t - t_i)^2\sigma_1\sigma - \frac{1}{6}(t - t_i)^3\sigma^2.$$

By summing up this yields the value of $m_G(t)$. The calculation of the variance involves several steps. First since the Brownian motions W_1^t and W_2^t are independent, we can write

$$V_{P^t}[\ln G(t)] = \frac{1}{n^2} V_{P^t}\left[\sum_{i=0}^{n-1} \int_{t_i}^{t} \sigma_2 dW_2^t(u)\right]$$

$$+ \frac{1}{n^2} V_{P^t}\left[\sum_{i=0}^{n-1}\left((t - t_i)\sigma W_1^t(t_i) + \int_{t_i}^{t} (\sigma_1 - (t - u)\sigma) dW_1^t(u)\right)\right].$$

The first variance term is the given by

$$V_{P^t}\left[\sum_{i=0}^{n-1} \int_{t_i}^{t} \sigma_2 dW_2^t(u)\right] = V_{P^t}\left[\sum_{i=0}^{n-1} (i + 1) \int_{t_i}^{t} \sigma_2 dW_2^t(u)\right] = \sum_{i=0}^{n-1}(i + 1)^2 \sigma_2^2 (t_{i+1} - t_i),$$

whereas the calculation of the second term is more complicated. Note that for $W_1^t(t_0) = 0$ we have

$$\sum_{i=0}^{n-1} (t - t_i)\sigma W_1^t(t_i) = \sum_{i=0}^{n-1} (t - t_i)\sigma W_1^t(t_0) + \sum_{i=0}^{n-1}\left(\sum_{j=i}^{n-1} (t - t_j)\right)\sigma(W_1^t(t_i) - W_1^t(t_{i-1}))$$

$$= \sum_{i=0}^{n-2}\left(\sum_{j=i+1}^{n-1} (t - t_j)\right)\sigma(W_1^t(t_{i+1}) - W_1^t(t_i)).$$

Furthermore we know that

$$\sum_{j=i+1}^{n-1} (t - t_j) = \sum_{j=i+1}^{n-1} (j - i)(t_{j+1} - t_j),$$

which implies that

$$\sum_{i=0}^{n-1} (t - t_i)\sigma W_1^t(t_i) = \sum_{i=0}^{n-2}\left[\sum_{j=i+1}^{n-1} (j - i)(t_{j+1} - t_j)\right]\sigma(W_1^t(t_{i+1}) - W_1^t(t_i)).$$

For the second variance term we therefore can conclude

$$V_{P^t}\left[\sum_{i=0}^{n-1}\left((t - t_i)\sigma W_1^t(t_i) + \int_{t_i}^{t} (\sigma_1 - (t - u)\sigma) dW_1^t(u)\right)\right]$$

$$= V_{P^t}\left[\sum_{i=0}^{n-2}\left(\sum_{j=i+1}^{n-1}(j-i)(t_{j+1}-t_j)\right)\sigma(W_1^t(t_{i+1})-W_1^t(t_i))\right.$$

$$\left.+\sum_{i=0}^{n-1}(i+1)\int_{t_i}^{t_{i+1}}(\sigma_1-(t-u)\sigma)dW_1^t(u)\right]$$

$$= V_{P^t}\left[\sum_{i=0}^{n-2}\int_{t_i}^{t_{i+1}}\left(\sum_{j=i+1}^{n-1}(j-i)(t_{j+1}-t_j)\right)\sigma+(i+1)(\sigma_1-(t-u)\sigma)dW_1^t(u)\right.$$

$$\left.+n\int_{t_{n-1}}^{t_n}(\sigma_1-(t-u)\sigma)dW_1^t(u)\right]$$

$$= \sum_{i=0}^{n-2}\int_{t_i}^{t_{i+1}}\left[\left(\sum_{j=i+1}^{n-1}(j-i)(t_{j+1}-t_j)\sigma\right)+(i+1)(\sigma_1-(t-u)\sigma)\right]^2du$$

$$+n^2\int_{t_{n-1}}^{t}(\sigma_1-(t-u)\sigma)^2du.$$

This implies that

$n^2V_{P^t}[\ln G(t)]$

$$= \sum_{i=0}^{n-1}[(i+1)^2(\sigma_1^2+\sigma_2^2)(t_{i+1}-t_i)]$$

$$+\sum_{i=0}^{n-1}\left[(i+1)^2\left[\sigma_1\sigma((t-t_{i+1})^2-(t-t_i)^2)-\frac{\sigma^2}{3}((t-t_{i+1})^3-(t-t_i)^3)\right]\right]$$

$$+\sum_{i=0}^{n-2}\left[\left(\sum_{j=i+1}^{n-1}(j-i)(t_{j+1}-t_j)\right)^2\sigma^2(t_{i+1}-t_i)\right]$$

$$+2\sum_{i=0}^{n-2}\left[\left(\sum_{j=i+1}^{n-1}(j-i)(t_{j+1}-t_j)\right)(i+1)\sigma\sigma_1(t_{i+1}-t_i)\right]$$

$$+\sum_{i=0}^{n-2}\left[\left(\sum_{j=i+1}^{n-1}(j-i)(t_{j+1}-t_j)\right)(i+1)\sigma^2[(t-t_{i+1})^2-(t-t_i)^2]\right].$$

We now assume that $t = t_{n^*(t)-1} + \Delta t = n^*(t) \cdot \Delta t$ where for simplicity we again set $n := n^*(t)$. With $E_{P^t}[G(t)] = \exp\{E_{P^t}[\ln G(t)] + 1/2\, V_{P^t}[\ln G(t)]\}$ and $\Sigma_{i=0}^{n-1}(t - t_i) = \Sigma_{i=0}^{n-1} (i + 1)(t_{i+1} - t_i)$ we can rewrite the equation for the expected geometric average as

$$\begin{aligned} E_{P^t}[G(t)] = & \left(\prod_{i=0}^{n-1} \frac{D(t_0, t_i)}{D(t_0, t)} \right)^{1/n} \\ & \cdot \exp\left\{ - \frac{\sigma_1^2 + \sigma_2^2}{2n^2} \sum_{i=0}^{n-1} (n - (i + 1))(i + 1)(t_{i+1} - t_i) \right\} \\ & \cdot \exp\left\{ + \frac{\sigma_1 \sigma}{2n^2} \sum_{i=0}^{n-1} [n(t - t_i)^2 - (i + 1)^2 (t_{i+1} - t_i)[2t - t_{i+1} - t_i]] \right\} \\ & \cdot \exp\left\{ + \frac{\sigma_1 \sigma}{2n^2} \sum_{i=0}^{n-2} \left[\left(\sum_{j=i+1}^{n-1} (j - i)(t_{j+1} - t_j) \right) (i + 1)(t_{i+1} - t_i) \right] \right\} \\ & \cdot \exp\left\{ - \frac{\sigma^2}{2n^2} \sum_{i=0}^{n-1} \left[n t_i (t - t_i)^2 + \frac{1}{3} n (t - t_i)^3 + \frac{1}{3} (i + 1)^2 ((t - t_{i+1})^3 \right.\right. \\ & \qquad \left.\left. - (t - t_i)^3) \right] \right\} \\ & \cdot \exp\left\{ + \frac{\sigma^2}{2n^2} \sum_{i=0}^{n-2} \left[\left(\sum_{j=i+1}^{n-1} (j - i)(t_{j+1} - t_j) \right)^2 (t_{i+1} - t_i) \right] \right\} \\ & \cdot \exp\left\{ - \frac{\sigma^2}{2n^2} \sum_{i=0}^{n-2} \left[\left(\sum_{j=i+1}^{n-1} (j - i)(t_{j+1} - t_j) \right) (i + 1)(t_{i+1} - t_i) \right.\right. \\ & \qquad \left.\left. [2t - t_{i+1} - t_i] \right] \right\}. \end{aligned}$$

Note first that

$$\left(\prod_{i=0}^{n-1} e^{r(t-t_i)} \right)^{1/n} = \exp\left\{ \frac{rt}{2} \frac{n(n + 1)}{n^2} \right\}.$$

The remaining sums are calculated as follows:

$$\frac{1}{n^2} \sum_{i=0}^{n-1} [(n - (i + 1))(i + 1)] \Delta t = \frac{t}{6} \frac{n^2 - 1}{n^2},$$

$$\frac{1}{n^2}\sum_{i=0}^{n-1}[n(n-i)^2-(i+1)^2(2n-2i-1)]\Delta t^2 = \frac{1}{6}\frac{(n-1)(n+1)^2}{n}\Delta t^2,$$

$$\frac{1}{n^2}\sum_{i=0}^{n-2}\left[\left(\sum_{j=i+1}^{n-1}(j-i)\right)(i+1)\right]\Delta t^2 = \frac{\Delta t^2}{24}\frac{n-1}{n}[n^2+3n+2],$$

$$\frac{1}{n^2}\sum_{i=0}^{n-2}\left[\left(\sum_{j=i+1}^{n-1}(j-i)\right)^2\right]\Delta t^3 = \frac{\Delta t^3}{60}\frac{n+1}{n}[3n^3-3n^2-2n+2],$$

$$\frac{1}{n^2}\sum_{i=0}^{n-2}\left[\left(\sum_{j=i+1}^{n-1}(j-i)\right)(i+1)(2n-2i-1)\right]\Delta t^3 = \frac{\Delta t^3}{40}\frac{n+1}{n}[2n^3+3n^2-3n-2],$$

and finally

$$\frac{1}{n^2}\sum_{i=0}^{n-1}\left[nt_i(t-t_i)^2+\frac{1}{3}n(t-t_i)^3+\frac{1}{3}(i+1)^2((t-t_{i+1})^3-(t-t_i)^3)\right]$$

$$= \frac{\Delta t^3}{n\cdot 180}[24n^4+15n^3-20n^2-15n-3].$$

Acknowledgments

Financial support from the Danish Natural and Social Science Research Council for the first author and from the Deutsche Forschungsgemeinschaft Sonderforschungsbereich 303 at the University of Bonn for both authors is gratefully acknowledged. Special thanks are due to Ruediger Frey for helpful discussions. The plots were produced using GNUPLOT 3.5.

Notes

1. As an alternative pure endowment or term contracts could have been analyzed. The methodology developed in this article is valid for these contracts as well.
2. The analysis covers the situation with a single initial premium payment. This case is much easier to solve than the periodic premium case as the path dependence introduced through the periodic investment in the mutual fund is eliminated.
3. No administrative costs are included in the analysis.
4. More precisely, we assume that the death or survival distribution is stochastically independent from the distribution of the mutual fund and the zero coupon bonds under the forward risk-adjusted measure that will be introduced in this section.
5. Alternatively, we could include so-called loading factors—that is, consider different survival and death distributions. This would not change our theoretical analysis, but the quantitative results on which we report in Section 6 would indicate higher premium values.

6. The notation will also be helpful in several of the proofs to follow.
7. Note that the function $1/\hat{x}(\cdot)$ is convex in b.
8. For more details see Nielsen and Sandmann [1995]. The generalized algorithms for the first four central moments are also given there.
9. Using the antithetic technique this yields in total $2 \cdot M$ paths.

References

AASE, K.K., and PERSSON, S. [1994]: "Pricing of Unit-Linked Life Insurance Policies," *Scandinavian Actuarial Journal*, 1, 26–52.

BACINELLO, A.R., and ORTU, F. [1993a]: "Pricing Equity-Linked Life Insurance with Endogenous Minimum Guarantees," *Insurance: Mathematics and Economics*, 12, 245–257.

BACINELLO, A.R., and ORTU, F. [1993b]: "Pricing Guaranteed Securities–Linked Life Insurance Under Interest-Rate Risk," *Actuarial Approach for Financial Risks*, 35–55.

BACINELLO, A.R., and ORTU, F. [1994]: "Single and Periodic Premiums for Guaranteed Equity-Linked Life Insurance Under Interest-Rate Risk: The 'Log-normal + Vasicek' Case," in *Financial Modelling*, L. Peccati and M.Viren (Eds.), Physica-Verlag, Heidelberg.

BRENNAN, M.J., and SCHWARTZ, E. [1976]: "The Pricing of Equity-Linked Life Insurance Policies with an Asset Value Guarantee," *Journal of Financial Economics*, 3, 195–213.

BRENNAN, M.J., and SCHWARTZ, E. [1979]: "Pricing and Investment Strategies for Equity-Linked Life Insurance," *Huebner Foundation Monograph 7*, Wharton School, University of Pennsylvania, Philadelphia.

DELBAEN, F. [1990]: "Equity-Linked Policies," *Bulletin Association Royal Actuaires Belges*, 33–52.

EKERN, S., and PERSSON, S.-A. [1996]: "Exotic Unit-Linked Life Insurance Contracts," *The Geneva Papers on Risk and Insurance Theory*, 21, 35–63.

EL KAROUI, N., LEPAGE, C., MYNENI, R., ROSEAU, N., and VISWANATHAN, R. [1991]: "The Valuation and Hedging of Contingent Claims with Markovian Interest Rate," Working paper, Université de Paris 6.

GEMAN, H., EL KAROUI, N., and ROCHET, J.-C. [1995]: "Changes of Numeraire, Changes of Probability Measure and Option Pricing," *Journal of Applied Probability*, 32, 443–458.

HO, T.S., and LEE, S.-B. [1986]: "Term Structure Movements and Pricing Interest Rate Contingent Claims," *Journal of Finance*, 41, 1011–1029.

JAMSHIDIAN, F. [1989]: "An Exact Bond Option Formula," *Journal of Finance*, 44, 205–209.

JAMSHIDIAN, F. [1991]: "Bond and Option Evaluation in the Gaussian Interest Rate Model," *Research in Finance*, 9, 131–170.

NIELSEN, J., and SANDMANN, K. [1995]: "Equity-Linked Life Insurance: A Model with Stochastic Interest Rates," *Insurance, Mathematics and Economics*, 16, 225–253.

PERSSON, S.A. [1994]: "Pricing Life Insurance Contracts Under Financial Uncertainty," Doctoral dissertation, Norwegian School of Economics and Business Administration, Bergen.

TURNBULL, S.M., and WAKEMAN, L.M. [1991]: "A Quick Algorithm for Pricing European Average Options," *Journal of Financial and Quantitative Analysis*, 26, 377–389.

VORST, T. [1992]: "Prices and Hedge Ratios of Average Exchange Rate Options," *International Review of Financial Analysis*, 1, 179–193.

The Geneva Papers on Risk and Insurance Theory, 21: 103–122 (1996)

Optimal Dynamic Hedging in Incomplete Futures Markets

ABRAHAM LIOUI
University of Paris-I Sorbonne and Bar Ilan University

PASCAL NGUYEN DUC TRONG
ESSEC and Université du Littoral

PATRICE PONCET
University of Paris-I Sorbonne and ESSEC, Département Finance, Avenue Bernard Hirsch, B.P. 105, 95021 Cergy-Pontoise Cedex, France

Abstract

This article derives optimal hedging demands for futures contracts from an investor who cannot freely trade his portfolio of primitive assets in the context of either a CARA or a logarithmic utility function. Existing futures contracts are not numerous enough to complete the market. In addition, in the case of CARA, the nonnegativity constraint on wealth is binding, and the optimal hedging demands are not identical to those that would be derived if the constraint were ignored. Fictitiously completing the market, we can characterize the optimal hedging demands for futures contracts. Closed-form solutions exist in the logarithmic case but not in the CARA case, since then a put (insurance) written on his wealth is implicitly bought by the investor. Although solutions are formally similar to those that obtain under complete markets, incompleteness leads in fact to second-best optima.

Key words: hedging, incomplete markets, use of futures, wealth constraint, exponential utility, logarithmic utility.

1. Introduction

The theory of dynamic hedging considers an investor holding a fixed, nontraded, portfolio of spot market assets and choosing a continuously rebalanced (hedge) portfolio of futures contracts so as to maximize the expected utility of his terminal wealth. Hedging profits and losses on the futures positions being marked to market in an interest-bearing (or -paying) margin account, terminal wealth is the sum of the final value of the fixed portfolio of spot market assets and the terminal value of the margin account. From an economic perspective, hedging is here akin to insurance in that both intend to improve, in an expected-utility sense, the distribution of future wealth.

Solutions to this hedging problem have been derived mostly in terms of the derivatives of the value function deemed to solve the Hamilton-Jacobi-Bellman equation for optimal control, which precludes closed-form expressions for the individual's various demand functions (see, for example, Anderson and Danthine [1983], Breeden [1984], Ho [1984], Stulz [1984], Adler and Detemple [1988a, 1988b], Poncet and Portait [1993], and Svensson and Werner [1993]). A notable exception is Duffie and Jackson [1990], who provide explicit solutions in several special cases. Unfortunately, in all cases except when the utility function exhibits infinite marginal utility at zero wealth (such as the logarithmic), the nonnegativity constraint on wealth is binding, and the space of admissible trading strategies is thus restricted. This makes the dynamic programming approach even more intricate (see Karatzas, Lehoczky, and Shreve [1987].

Lioui and Poncet [1996] assuming that utility functions exhibit constant absolute risk aversion (CARA) and that asset price increments are lognormal, characterize the optimal hedging demands for futures contracts when the nonnegativity constraint on wealth is introduced and show that the "speculative" component of the latter is smaller, in absolute terms, than is usually reported. To achieve this, they apply the Karatzas, Lehoczky, and Shreve [1987] and Cox and Huang [1989, 1991] methodology to transform the dynamic optimal control problem into a static one and to do so assume that markets are complete.

This directly raises the even more crucial issue of market incompleteness. Markets are incomplete when there may exist derivative instruments written on the traded assets that cannot be replicated by dynamically trading on these assets. Hence some risks remain genuinely unhedgeable. A necessary condition for incompleteness is that the number of traded risky securities N is strictly smaller than the dimension L of the Brownian motion driving the market prices.

All previously quoted works but Duffie and Jackson [1990] and the end part of Adler and Detemple [1988b] assume completeness, and its relaxation was long awaited until significant advances relative to optimal consumption and portfolio decisions were made by He and Pearson [1991] and Karatzas, Lehoczky, Shreve, and Xu [1991], following a pioneering step by Pagès [1987].[1] Adler and Detemple [1988b] examine the issue of optimal hedging by a Bernoulli investor in both cases of perfect and imperfect correlation between the spot asset and its associated futures. In their setting, those cases are equivalent to complete and incomplete markets, respectively. In the latter case, even though the investor has a logarithmic utility, they cannot provide an entirely explicit solution for his hedging demand for futures. In addition, they incorrectly claim that this demand will include a Merton-Breeden component—that is, a hedge against unfavorable shifts in the opportunity set. Thus even when utility is logarithmic and prices are lognornally distributed or more generally follow diffusion processes, there is no known solution to the optimal hedging problem under incomplete markets. One of the special cases studied by Duffie and Jackson [1990] is when utility functions are CARA. Although they provide an explicit solution and are even able to characterize the hedging demands at equilibrium under various assumptions, which is of special interest under incomplete markets, they do not take into account the (binding) nonnegativity constraint on wealth, and they assume that asset-price increments are Gaussian rather than lognormal. In addition to allowing negative asset prices, this assumption also permits hedgers' myopia, by which is meant that investors hedge only local changes in wealth and thus construct optimal dynamic hedges that are the same as the corresponding static ones.

This article intends (1) to solve explicitly the optimal hedging problem in the logarithmic utility case under genuinely incomplete markets and, in particular, imperfect correlation between spot assets and futures contracts and (2) to solve the same problem in the CARA utility case when price increments are lognormal and the nonnegativity constraint on wealth is explicitly taken into account.

A disclaimer may be useful at this stage. In spite of the generality implied by its title, this article does not attempt to address the hedging problem faced by a firm or a corporation. It is well known from Modigliani-Miller's work that under perfect markets, corporate hedging (and insurance) is irrelevant since shareholders hold diversified portfolios in which the risk-return tradeoff is optimized. In actual practice, however, markets being both imperfect and incomplete, firms do engage in various hedging activities. An appropriate framework to address this issue is in need, in order to assess whether and to what extent firms should hedge.

The remainder of the article is organized as follows. Section 2 outlines our framework, while Section 3 describes the methodology adopted. Section 4 derives the dynamics of optimal wealth and the optimal hedging demands for futures contracts under incomplete markets in the logarithmic case. Section 5 analyzes the more complex case of the CARA utility function. Section 6 contains conclusions and offers a brief discussion of possible directions for further research. An appendix includes one technical derivation.

2. The framework

We consider an individual choosing a dynamic futures trading strategy so as to maximize his expected utility of wealth at a future date T, in the following environment. This set of assumptions holds for the logarithmic case. Some assumptions will be modified later on for dealing with CARA utilities.

Assumption 1. *There exist M primitive assets to be hedged, whose value obeys an M-dimensional diffusion process S supposed to be well defined, with stochastic differential representation:*

$$dS_t = I_{S_t}\mu(t, S_t)dt + I_{S_t}\sigma(t, S_t)dZ_t, \tag{1}$$

where I_{S_t} *denotes the diagonal* $(M \times M)$ *matrix with elements* S_t^j $(j = 1, \ldots, M)$, μ *is an M-dimensional process,* σ *is an* $(M \times L)$ *matrix, and Z denotes a standard Brownian motion in* R^L, $Z = (Z^1 \ldots, Z^L)^\tau$, *where* τ *indicates transpose. Hence the uncertainty is formalized by the complete filtered space* $(\Omega, \mathbf{F}, \{\mathbf{F}_t\}_{t\in[0,T]}, P)$, *where P is the actual (historical) probability, and the filtration is the augmented filtration generated by the Brownian motion.*

We posit $M < L$. Although the matrix $\sigma(\cdot)$ is assumed to be of full rank, primitive markets thus are not strictly complete even for unconstrained investors—that is, those who can freely trade on these assets.[2]

To ease the exposition, we note $\mu(t, S_t) = \mu_t$ and $\sigma(t, S_t) = \sigma_t$. Also, the primitive assets do not pay any dividends between 0 and T.

Assumption 2. *There exists a savings account B(t) following the process*

$$dB_t = r(S_t, t)\, B_t\, dt, \text{ with } B(0) = 1, \tag{2}$$

where $r(\cdot)$ *is a nonnegative process. The savings account is locally riskless and earns the riskless rate. Investors may accumulate a positive or a negative position on this account. To simplify, we note* $r(S_t, t) = r_t$.

Assumption 3. *There exist N futures contracts available for trade, whose maturity date, T, coincides with the investor's horizon.*[3] *We assume $N < M < L$: there are fewer futures than primitive asset, and each futures contract is not necessarily written on one particular asset.*[4] *The futures prices process F_t is given by:*

$$dF_t = I_{F_t}\, \eta(t, F_t)dt + I_{F_t}\, \nu(t, F_t)dZ_t, \tag{3}$$

where I_{F_t} denotes the diagonal $(N \times N)$ matrix with the elements F_t^j $(j = 1, \ldots, N)$, η is an N-dimensional process, and v is an $(N \times L)$ matrix of full rank. To ease the exposition, we note $\eta(t, F_t) = \eta_t$ and $\nu(t, F_t) = \nu_t$.[5]

The fact that each futures contract is not necessarily written on a particular asset allows us to introduce, in a natural and convenient manner, the possible benefits of cross-hedging. This is specially relevant and useful in an international portfolio context. For instance, the domestic investor may essentially invest domestically in primitive assets but, due to transaction or information costs, find it more appropriate to use futures instruments as foreign hedging vehicles.

Notice that the existence of these $N < M$ futures contracts is not enough to ensure the completeness of the market for the constrained investor who can not freely trade the primitive assets S, even though M were equal to L. Consequently, the filtration generated by the futures, F^E, is smaller than the filtration generated by the primitive assets, F^S, itself smaller than F.

A hedge is an N-dimensional stochastic process $\theta \equiv (\theta^1, \ldots, \theta^N)^T$ specified by a futures position in each futures contract.[6] At time t, the position is marked to market, so that gains and losses generated by futures price changes accrue to the hedger's margin account in a continuous manner.

Assumption 4. *Markets are frictionless, and there are no arbitrage opportunities. This implies that there exists at least a probability measure, associated with the locally riskless asset B_t, that makes discounted futures prices martingales.*[7]

From Pagès [1987], we know that any F_t-adapted process ξ, that satisfies

$$\eta_t - r_t \mathbf{1} = \nu_t \xi_t, \quad \forall t \in [0, T]. \tag{4}$$

P, almost surely, is a possible market price of risk for the constrained investor.[8] Then the measure Q defined by its Radon-Nikodym derivative with respect to P:

$$\left.\frac{dQ}{dP}\right|_{F_T} = \exp\left\{-\int_0^T \xi_t^T dZ_t - \frac{1}{2}\int_0^T \|\xi_t\|^2\, dt\right\}$$

is one such equivalent martingale measure (see also He and Pearson [1991]), Proposition 1).

Processes ξ verifying (4) can best be viewed when separated into two components: the first one in the range of ν_t and the second one in the kernel of ν_t. As under complete markets, the component lying in the range of ν_t, at each time t, is unique. This should not be surprising since the restriction to traded risks makes the market necessarily complete. Denote κ this process. Then it is easy to show, by projection arguments, that

$$\kappa_t = \nu_t^T(\nu_t \nu_t^T)^{-1} (\eta_t - r_t 1). \tag{5}$$

In turn, $\xi = \kappa$ yields a remarkable probability measure that Föllmer and Schweizer [1991] have dubbed the minimal martingale measure.

As shown by Pagès [1987], the other equivalent martingale measures, noted Q^δ, are generated by processes δ in Ker(ν) defined by

$$\text{Ker}(\nu) \equiv \left\{ \delta\colon \int_0^T \|\delta_t\|^2 dt < \infty \text{ and } \forall t \in [0, T],\ \nu_t \delta_t = 0 \right\}.$$

Thus,

$$\left.\frac{dQ^\delta}{dP}\right|_{F_T} = \exp\left\{ -\int_0^T (\kappa_t^T + \delta_t^T) dZ_t - \frac{1}{2}\int_0^T \|\kappa_t + \delta_t\|^2 dt \right\}.$$

Note also that under Q^δ, using Girsanov's theorem, the process Z_t^δ such that $Z_t^\delta = Z_t + \int_0^t (\kappa_s + \delta_s) ds$ is a standard Brownian motion.

Finally, we will need the real process $H_t(\delta) = B_t(dQ^\delta/dP|_{F_t})^{-1}$, which is written explicitly

$$H_t(\delta) = \exp\left\{ \int_0^t \left(r_s + \frac{1}{2}\|\kappa_s + \delta_s\|^2 \right) ds + \int_0^t (\kappa_s^T + \delta_s^T) dZ_s \right\}. \tag{6}$$

$H_t(\delta)$ is the value at time t of the numéraire portfolio, which makes the H-denominated value process of any admissible portfolio a martingale under the historical probability measure P (see Long [1990], for instance). This numéraire portfolio, which is the optimal portfolio of the Bernoulli investor, plays an increasing role in theoretical work relative to dynamic investment and hedging strategies (see, for instance, Bajeux-Besnainou and Portait [1994] and Gourieroux, Laurent, and Pham [1995]).

Assumption 5. *The margin account current value, denoted by X_t, is credited or debited (if negative) with interest at the locally riskless rate r_t. The margin account then is equal to*

$$X_t = \int_0^t \frac{B_t}{B_s} \theta_s^T dF_s.$$

Applying Ito's lemma and using (2) *and* (3) *yields*

$$dX_t = r_t X_t dt + \theta_t^T dF_t = (r_t X_t + \theta_t^T I_{F_t} \eta_t) dt + \theta_t^T I_{F_t} \nu_t dZ_t$$

Assumption 6. *The investor holds a fixed, nontraded, portfolio* Π *of the M primitive assets, with terminal random value* $\Pi^T S_T$ *at horizon date T. Given the futures strategy* θ, *his total wealth at time t is* W_t *such that*

$$W_t = \Pi^T S_t + X_t,$$

which yields

$$\begin{aligned} dW_t &= \Pi^T dS_t + dX_t \\ &= (\Pi^T I_{S_t} \mu_t + r_t X_t + \theta_t^T I_{F_t} \eta_t)dt + (\Pi^T I_{S_t} \sigma_t + \theta_t^T I_{F_t} v_t)dZ_t. \end{aligned} \tag{7}$$

Two remarks may be in order here. First, allowing the investor to trade on some primitive assets would not change the nature or the scope of our analysis but only introduce useless algebra. Second, and more important, one might investigate the likely effects that hedging possibilities have on the investor's initial commitment to any particular portfolio Π, just before hedging actually takes place. Under complete markets, it is clear that irrespective of the particular Π given initially, the investor will attain his first-best optimum. Under incomplete markets, however, the choice of a particular Π will be affected by the knowledge that hedging possibilities exist or not. In accordance with the literature on hedging, we implicitly assume that our investor's fixed portfolio Π has been selected optimally, given that hedging is partially available.

Assumption 7. *The investor is a Von Neumann-Morgenstern expected utility of terminal wealth maximizer and his utility function is logarithmic—that is,* $U(W_T) = LnW_T$, *his relative risk aversion coefficient being the constant one.*

3. The methodology

We explain briefly here the general ideas behind the methodology adopted here, inspired from that of He and Pearson [1991], Cox and Huang [1989, 1991], and Karatzas et al. [1991]. For clarity of exposition, we suppose first that markets are complete and then discuss the case of incomplete markets.

When markets are complete, the main idea is to transform the dynamic optimal control problem into a static one by solving first for the terminal optimal wealth W_T. Then we compute the value W_t, at each time t, of this wealth. Identifying W_t with the wealth process given by (7), it is easy to get the optimal hedging demands for futures contracts. When markets are complete, it is clear that there always exists a replicating strategy using futures that insures the chosen level of this terminal wealth.

In the process, two cases are possible. For some utility functions, the nonnegativity constraint on wealth is not binding, such as when the utility function exhibits infinite marginal utility at zero wealth (e.g., the logarithmic, our Assumption 7). For the others (such as our Assumption 7′ further on) the constraint is binding, but Cox and Huang [1989, 1991] provide us with the following powerful equivalence result: the solution to the constrained program is equal to the solution to the unconstrained one to which is added an insurance contract (a protective put of strike zero written on the unconstrained terminal wealth) that will pay off the negative wealth if it occurs. In other words, the optimal constrained strategy will allocate the entire initial wealth between the unconstrained strategy and the protective put written on the unconstrained strategy. This result will be used in this article, since it holds whether one deals with a portfolio choice problem, as Cox and Huang did, or with a hedging problem, as we do.

When markets are incomplete, the portfolio decision problem is more involved for two reasons. First, there are as many budget constraints as there are martingale measures associated with the chosen numéraire—that is, an infinity. Second, even if the static problem has a solution, the resulting optimal wealth is not necessarily replicable. However, as shown in the discussion following Assumption 4, we know that there exists a subspace of risk premia processes on the risks uncorrelated with the traded risks (to simplify, "the subset of uncorrelated risks"). To each process of this subspace corresponds an equivalent martingale measure for which the rates of return on all traded assets are equal to the riskless rate.

The solution requires three steps. First, a vector belonging to the kernel of the diffusion matrix is arbitrarily selected and the investor's static program is solved (subject to a budget constraint that makes use of the probability measure determined by this vector), the solution depending on the vector chosen. In the second step, one computes the value function of this program—that is, the investor's expected utility obtained with this optimal terminal wealth. In order for a replication strategy for this wealth to exist, it is necessary and sufficient that the selected vector of the kernel minimizes the investor's value function. The dual program of the static (primal) one is constructed, the solution to which is a vector belonging to the kernel and thus a martingale measure. Using He and Pearson's [1991] terminology, it is a minimax martingale measure. Two possible interpretations of this finding are offered. According to He and Pearson [1991], the investor faces the worst situation (the value function is minimal) due to the market incompleteness. According to Karatzas et al. [1991], the vector arbitrarily chosen in the first step can be interpreted as the vector of equilibrium risk premia attached to the fictitious assets that complete the investor's market. However, real optimal wealth is replicable only if those assets are not present in the investor's replication strategy. To achieve this, these risk premia must be set to a level such that it is suboptimal for the individual to invest in these fictitious assets. Finally, the third step requires that the terminal wealth replication strategy be constructed by using the martingale measure associated with the particular vector of the kernel that solves the investor's dual program. The Cox and Huang program constructed with this measure thus leads to the optimal solution and the optimal strategy using only the traded assets.

When solving an optimal hedging problem, however, we cannot directly apply the above method since the demand for non traded assets is fixed and strictly positive (we call it exogenous) and cannot be hedged by the existing futures in a genuinely incomplete market. If the market were complete as above, one knows from Merton [1971, 1973] that the investor's optimal portfolio would comprise the tangent portfolio and those portfolios that are most correlated with the sources of risk.[9]

The exact method we use here, under incompleteness of markets, consists in choosing the particular vector of the kernel in such a way that the investor's demand for the tangent portfolio and the most correlated portfolios coincide, on the subspace of uncorrelated risks, with the exogenous demand implicit in his fixed portfolio. It turns out that this particular vector also stems from a minimax procedure à la He and Pearson [1991] but where the optimal terminal distribution is implied by the fixed position on the primitive assets and the optimal position on the futures contracts.

We use a three-step procedure. In the first step, we show the equivalence between the dynamic program (DP) and a static program (SP_δ) where the market is fictitiously completed. In the second step, we solve (SP_δ) and choose the particular vector δ_t^* of the kernel Ker(ν) such that the derived demands are equal to the exogenous demands. In the third step, we verify that the solution derived with δ_t^* is indeed attainable—that is, that the component of optimal terminal wealth due to trading on futures is indeed F_T^F-measurable.

Since the first step is common to the logarithmic and the CARA functions, in fact to any utility function possessing the usual properties, we solve it now before dealing with the two specific cases.

The investor's initial dynamic program is written:

$$(DP)\quad \begin{cases} \underset{\theta \in \Theta}{\text{Max}}\ E^P U[\Pi^\tau S_T + X_T] \\ \text{s.t. } \Pi^\tau S_T + X_T \geq 0, \\ \text{and } dX_t = r_t X_t dt + \theta_t^\tau dF_t, \end{cases}$$

where θ belongs to Θ, the set of admissible, self-financing, hedging strategies.

Consider the seemingly unrelated static program

$$(SP_\delta)\quad \begin{cases} \underset{W_T \in L_+^2(F_T)}{\text{Max}}\ E^P U[W_T] \\ \text{s.t. } E^{Q^\delta}\left[\dfrac{W_T}{B_T}\right] = E^{Q^\delta}\left[\dfrac{\Pi^\tau S_T}{B_T}\right]. \end{cases}$$

In this program, the relevant space is L_+^2 (F_T), the space of all positive square integrable, F_T-measurable distributions of total terminal wealth, while in (DP) one considers only the distributions $\theta_T^\tau F_T$ of F_T^F, to which is added the distribution $\Pi^\tau S_T \in L^2(F_T^S)$.

For our purposes, it is sufficient to show that the program (SP_δ) implies the program (DP), although it is straightforward to show the converse—that is, that the two programs are equivalent. Suppose that $W_T^*(\delta)$ is the optimal solution to (SP_δ). Then $W_T^*(\delta) - \Pi^\tau S_T$ is solution to the following program:

$$(SP_\delta')\quad \begin{cases} \underset{X_T \in L^2(F_T^F)}{\text{Max}}\ E^P U[\Pi^\tau S_T + X_T] \\ \text{s.t. } \Pi^\tau S_T + X_T \geq 0, \\ \text{and } E^{Q^\delta}\left[\dfrac{X_T}{B_T}\right] = 0, \end{cases}$$

provided $W_T^*(\delta) - \Pi^\tau S_T \in L^2(F_T^F)$, which stresses the necessary measurability of the hedging process with respect to the filtration generated by the futures process.

But in (SP_δ'), the constraint

$$E^{Q^\delta}\left[\frac{X_T}{B_T}\right] = 0$$

is verified for all processes δ in Ker(ν) since X_T is F_T^F-measurable. In turn, this means that X_T is attainable by means of a futures trading strategy with zero initial wealth. Therefore X_T is to be taken from the set of distributions defining program (DP), and (SP_δ) implies (DP) as claimed.

4. The logarithmic investor's case

When the investor has a logarithmic utility function, we have readily the solution to program (SP_δ) with $U(W) = Ln\ W$:

$$W_T^*(\delta) = W_0(\delta)H_T(\delta),$$

where $W_0(\delta) \equiv E^{Q^\delta}[\Pi^\tau S_T/B_T]$ and H_T is defined by (6).

The optimal wealth process starting with endowment $W_0(\delta)$ and reaching $W_T^*(\delta) = W_0(\delta)H_T(\delta)$ almost surely is the martingale under Q^δ defined by

$$W_t^* = B_t E^{Q^\delta}\left[\frac{W_T^*}{B_T} \mid F_t\right].$$

Taking the expectation under probability P, we get

$$W_t^* = H_t(\delta) E^P\left[\frac{W_T^*}{H_T(\delta)} \mid F_t\right].$$

Thus, $W_t^*(\delta) = W_0(\delta)H_t(\delta)$, and the dynamics of optimal wealth is given by

$$\begin{aligned} dW_t^*(\delta) &= W_0(\delta)dH_t(\delta) \\ &= W_t^*(r_t + \|\kappa_t + \delta_t\|^2)dt + W_t^*(\kappa_t + \delta_t)^\tau dZ_t. \end{aligned} \tag{8}$$

We now determine under which conditions the process $dW_t^*(\delta)$ can be replicated by the appropriate trading on futures contracts. For so doing, one needs to identify the current wealth process (7) with the optimal process (8). Notice that it is sufficient to identify the diffusion coefficients, forgetting altogether the drift terms, since by arbitrage two optimal strategies of identical initial value and identical risk must have the same expected return. We thus obtain

$$W_t^*(\delta)\kappa_t^\tau + \delta_t^\tau) = \Pi^\tau I_{S_t}\sigma_t + \theta_t^\tau I_{F_t}\nu_t.$$

This equation can be separated in R^L along the two orthogonal spaces Range(v_t) and Ker(v_t), which yields, respectively,

$$W_t^*(\delta)\kappa_t^T = \Pi^T I_{S_t}\sigma_t v_t^T(v_t v_t^T)^{-1} v_t + \theta_t^T I_{F_t} v_t \tag{9}$$

$$W_t^*(\delta)\delta_t^T = \Pi^T I_{S_t}\sigma_t(I_L - v_t^T(v_t v_t^T)^{-1} v_t), \tag{10}$$

where I_L stands for the identity ($L \times L$) matrix.

It thus appears that once equation (10) has been verified by the appropriate choice of δ_t^*, we are left to determine the optimal trading strategy θ_t^* that validates (9).

Therefore, we first focus on equation (10). The LHS contains the optimal demand for uncorrelated risk (priced by δ_t). For the Bernoulli investor, this demand reduces to the tangent portfolio. Rubinstein [1976] has shown that the logarithmic utility is the relevant "benchmark" utility in Finance due to its myopic property: in a multiperiod context, as here, it is not the linear utility individual who is risk neutral—that is, who does not take into account shifts in his investment opportunity set—but the Bernoulli investor. Only an investor more risk averse than the logarithmic individual will be willing to pay an (insurance) premium to ensure a consumption level at a given future date. Due to this unique myopic property, the log-utility investor does not try to hedge unfavorable shifts in his opportunity set, and there will be no Merton-Breeden component in the hedging strategy. The RHS consists of the demands for uncorrelated risks implicit in the fixed position on nontraded assets.[10]

Given this price process δ_t^*, let us turn now to equation (9). The LHS states that the optimal demand for traded risks should be taken proportional to both optimal wealth W_t^* and "price" κ_t.[11] The RHS may already contain some traded risks given the fixed portfolio of primitive assets. However, there is little chance that this exogenous demand is equal to the optimal demand on the LHS. But this is precisely the role of futures contracts to bridge the gap between the optimal and the exogenous demands. Optimally, this trading strategy will satisfy

$$(\theta_t^*)^T I_{F_t} = -\Pi^T I_{S_t}\sigma_t v_t^T(v_t v_t^T)^{-1} + W_t^*(\eta_t - r_t 1)^T(v_t v_t^T)^{-1}, \tag{11}$$

where we have made use of expression (5) defining κ_t.

The first term in the RHS is the so-called pure hedge component of the strategy. This part of the overall position in futures tends to offset the nontraded cash position, up to a factor that has the dimension of a covariance (between cash and futures) over a variance (of futures price changes). Note however that we do not recover the usual result $-(B_t/B_T)\Pi^T I_{S_t}$ unless we assume that futures and primary assets are of equal number and perfectly correlated. In that particular case of perfect correlation, implying that markets are not genuinely incomplete, it is easy to show that $\delta^* = 0$—that is, the optimal solution is indeed formally the same as under complete markets (see Lioui and Poncet [1996]). Using $\sigma_t = v_t$ and $I_{F_t}v_t = (B_T/B_t)I_{S_t}\sigma_t$ from arbitrage between cash assets and their associated futures, equation (11) would read $\theta_t^* = -(B_t/B_T)\Pi + W_t^*(I_{F_t}v_t^T)^{-1}\kappa_t$.

The second term represents the "speculative" component. It depends on the absolute risk tolerance of the investor's value function, W_t^* (also his wealth), and on the expected return-variance tradeoff offered by the futures contracts. Thus it is a demand driven by mean-variance efficiency motives.

The remarkable feature of equation (11) is in fact that it holds in incomplete markets: in particular, futures contracts need not be perfectly correlated with their underlying cash instruments (they are not even necessarily written on one particular primitive asset and their number can be smaller than that of the cash assets). Thus, in contradiction with what Adler and Detemple [1988b] claimed, in the case of a logarithmic investor, we can derive an explicit solution for the optimal hedging demands for futures in a portfolio context when correlations between cash and futures instruments are imperfect. In particular, because of the Bernoulli investor's myopia, there are no Merton-Breeden hedging components. This contributes to the bulk of results in financial economics that are based on the properties of the logarithmic utility function.

To conclude, we can be sure, by choosing the δ_t^* that verifies (10), that the distribution W_t^* is indeed replicable by the trading policy $\theta_t^*(\delta^*)$ of equation (11). Therefore, $W_T^*(\delta^*) - \Pi^\tau S_T$ is F_T^F-measurable, and as seen in Section 3, $\theta_t^*(\delta^*)$ is the optimal control of the dynamic program (*DP*).

In order to provide the reader with additional insight, we offer below a very simple illustrative numerical example. Suppose, as a special case, that $L = 3$ (sources of risk), $M = 2$ (primitive assets), and $N = 1$ (futures contract). Let the investor's initial wealth be \$100,000, Π^τ be (1200 1000) the number of shares held in assets 1 and 2, respectively, $S^1 = \$50$ and $S^2 = \$40$ be the initial market prices of assets 1 and 2, respectively, r be 10%, the expected return on the futures contract be $\eta = 12\%$, the v (1 × 3) diffusion matrix for the futures be [0.14 0.22 0.12], and the σ (2 × 3) diffusion matrix for the spot assets be

$$\begin{bmatrix} 0.10 & 0.15 & 0.20 \\ 0.12 & 0.30 & 0.08 \end{bmatrix}.$$

Thus our Bernoulli investor holds initially \$60,000 in asset 1 and \$40,000 in asset 2. The first (hedge) component of his futures position (see equation (11)) is equal to

$$-\Pi^\tau I_{S_t}\sigma_t v_t^\tau (v_t v_t^\tau)^{-1} = -[1200 \quad 1000] \begin{bmatrix} 50 & 0 \\ 0 & 40 \end{bmatrix} \begin{bmatrix} .10 & .15 & .20 \\ .12 & .30 & .08 \end{bmatrix} \begin{bmatrix} .14 \\ .22 \\ .12 \end{bmatrix} \frac{1}{.0824}$$

$$= -\$96{,}553,$$

which, due to imperfect correlation, is not equal to −\$100,000.

The "speculative" component of equation (11) is equal to

$$W_t^*(\eta_t - r_t 1)^\tau (v_t v_t^\tau)^{-1} = 100{,}000\ (0.02)\ \frac{1}{.0824} = \$24{,}272,$$

which is positive because the expected return on the futures contract in excess of the riskless interest rate is here positive (12% − 10% = 2%). The investor's overall position in futures thus is −\$72,281.

5. The CARA investor's case

As noted at the outset of Section 3, in order to have a tractable problem, we have to modify some assumptions and work in a less general setting. We list below only those assumptions that have been changed.

Assumption 1′ (primitive assets). *The functions* $\mu(\cdot)$ *and* $\sigma(\cdot)$ *in equation* (1) *are deterministic functions of time.*

Assumption 2′ (riskless asset). *The process* $r(\cdot)$ *is deterministic.*

Assumption 3′ (futures). *The functions* $\eta(\cdot)$ *and* $v(\cdot)$ *in equation* (3) *are deterministic functions of time.*

It then follows that the market prices of risk κ_t and δ_t are deterministic, though not necessarily constant.

Assumption 7′ (utility function). *The investor is a Von Neumann-Morgenstern expected utility of terminal wealth maximizer and his utility function exhibits constant absolute risk aversion, i.e.* $U(W_T) = -e^{-\gamma W_T}$ *where* $\gamma > 0$ *is the coefficient of absolute risk aversion.*

As explained in Section 3, we solve the static utility maximization program for a given premium process $\delta \in \mathrm{Ker}(v)$:

$$(SP_\delta)\quad \begin{cases} \underset{W_T \in L^2_+(F_T)}{\mathrm{Max}} \quad E^P[-e^{-\gamma W_T}] \\ \\ \text{s.t. } E^{Q^\delta}\left[\dfrac{W_T}{B_T}\right] = W_0. \end{cases}$$

Since the nonnegativity constraint on wealth is binding, as is the case with the exponential utility postulated now (Assumption 7′), we derive first the solution to the unconstrained program and then compute the optimal terminal wealth for the constrained problem, using the Cox and Huang [1989, 1991] method. We will denote $\hat{W}$ the solution to the unconstrained problem, with initial value $\hat{W}_0$, and denote W^* the solution to the constrained problem, with initial value W_0. Hence, $W_0 \geq \hat{W}_0$.

The investor's unconstrained program is written

$$(USP_\delta)\quad \begin{cases} \underset{\hat{W}_T \in L^2_+(F_T)}{\mathrm{Max}} \quad E^P[-e^{-\gamma \hat{W}_T}] \\ \\ \text{s.t. } E^{Q^\delta}\left[\dfrac{\hat{W}_T}{B_T}\right] = \hat{W}_0, \end{cases}$$

where $\hat{W}_0$ stands for the initial wealth necessary to solve this program. It will become clear that this amount is less than W_0. Notice that both W_0 and $\hat{W}_0$ depend on the choice of δ. However, we do not need to explicit this dependency until equation (20).

The unique solution to (USP_δ) is such that

$$\gamma e^{-\gamma \hat{W}_T} - \frac{\lambda}{H_T(\delta)} = 0,$$

where λ is the Lagrangian multiplier, or

$$\hat{W}_T = -\frac{1}{\gamma} \text{Ln}\left(\frac{\lambda}{\gamma}\right) + \frac{1}{\gamma} \text{Ln}[H_T(\delta)].$$

Substituting $H_T(\delta)$ for its integral expression, we get

$$\begin{aligned}\hat{W}_T &= -\frac{1}{\gamma} \text{Ln}\left(\frac{\lambda}{\gamma}\right) + \frac{1}{\gamma}\left[\int_0^T (\kappa_t^T + \delta_t^T)dZ_t + \int_0^T \left(r_t + \frac{1}{2}\|\kappa_t + \delta_t\|^2\right) dt\right] \\ &= -\frac{1}{\gamma} \text{Ln}\left(\frac{\lambda}{\gamma}\right) + \frac{1}{\gamma}\left[\int_0^T (\kappa_t^T + \delta_t^T)dZ_t^\delta + \int_0^T \left(r_t - \frac{1}{2}\|\kappa_t + \delta_t\|^2\right) dt\right] \\ &= -\frac{1}{\gamma} \text{Ln}\left(\frac{\lambda}{\gamma}\right) + \frac{1}{\gamma}\int_0^T \left(r_t - \frac{1}{2}\|\kappa_t + \delta_t\|^2\right) dt + \frac{1}{\gamma}\int_0^T (\kappa_t^T + \delta_t^T)dZ_t^\delta.\end{aligned}$$

Then the budget constraint $E^{Q^\delta}[\hat{W}_T/B_T] = \hat{W}_0$ writes

$$\begin{aligned}\hat{W}_0 B_T &= -\frac{1}{\gamma} \text{Ln}\left(\frac{\lambda}{\gamma}\right) + \frac{1}{\gamma}\int_0^T \left(r_t - \frac{1}{2}\|\kappa_t + \delta_t\|^2\right) dt + E^{Q^\delta}\left[\frac{1}{\gamma}\int_0^T (\kappa_t^T + \delta_t^T)dZ_t^\delta\right] \\ &= -\frac{1}{\gamma} \text{Ln}\left(\frac{\lambda}{\gamma}\right) + \frac{1}{\gamma}\int_0^T \left(r_t - \frac{1}{2}\|\kappa_t + \delta_t\|^2\right) dt.\end{aligned}$$

Substituting back into $\hat{W}_T$, to eliminate λ, we get

$$\hat{W}_T = \hat{W}_0 B_T + \frac{1}{\gamma}\int_0^T (\kappa_t^T + \delta_t^T)dZ_t^\delta,$$

where it is readily seen that terminal wealth can become negative.

From this equation, one can derive the value at time t of the optimal "unconstrained" wealth:

$$\hat{W}_t = \hat{W}_0 B_t + \frac{1}{\gamma}\frac{B_t}{B_T}\int_0^t (\kappa_s^T + \delta_s^T)dZ_s^\delta. \tag{12}$$

Given Cox and Huang, the solution to the "constrained" program (SP_δ) is

$$W_T^* = [\hat{W}_T]^+. \tag{13}$$

It is worth noting at this stage that

- Knowing the investor's actual wealth W_0 and the solution to the unconstrained program (10), we readily get the component $\hat{W}_0$ devoted to the unconstrained strategy, and
- $[\hat{W}_T]^+ = \hat{W}_T + \max[0; -\hat{W}_T]$, which highlights the put (insurance) component of the optimal constrained strategy.

In order to derive the value of optimal wealth at time t, define A the event of nonnegative unconstrained terminal wealth $\hat{W}_T$:

$$A \equiv \{\omega \in \Omega: \hat{W}_T \geq 0\},$$

and denote 1_A its indicator function.

The value W_t^* at time t of optimal terminal wealth given by (13) is such that

$$\frac{W_t^*}{B_t} = E^{Q^\delta}\left[\frac{W_T^*}{B_T}|F_t\right].$$

Hence, using (12)

$$\begin{aligned} W_t^* &= \frac{B_t}{B_T} E^{Q^\delta}\left[\left[\hat{W}_0 B_T + \frac{1}{\gamma}\int_0^T (\kappa_s^T + \delta_s^T) dZ_s^\delta\right]^+ |F_t\right] \\ &= \frac{B_t}{B_T} E^{Q^\delta}\left[\left(\hat{W}_0 B_T + \frac{1}{\gamma}\int_0^t (\kappa_s^T + \delta_s^T) dZ_s^\delta + \frac{1}{\gamma}\int_t^T (\kappa_s^T + \delta_s^T) dZ_s^\delta\right) 1_A | F_t\right] \\ &= \left(\hat{W}_0 B_t + \frac{1}{\gamma}\frac{B_t}{B_T}\int_0^t (\kappa_s^T + \delta_s^T) dZ_s^\delta\right) E^{Q^\delta}[1_A | F_t] \\ &\quad + \frac{1}{\gamma}\frac{B_t}{B_T} E^{Q^\delta}\left[\left(\frac{1}{\gamma}\int_t^T (\kappa_s^T + \delta_s^T) dZ_s^\delta\right) 1_A | F_t\right] \end{aligned} \quad (14)$$

Since our assumptions imply that the market prices of risk κ_t and δ_t are deterministic, we can compute explicitly the expectations present in the above equation.[12]

First, notice that the term multiplying $E^{Q^\delta}[1_A | F_t]$ in (14) is nothing but $\hat{W}_t$. Then, nonnegative terminal wealth $\hat{W}_T$ (event A) occurs when

$$\hat{W}_0 B_T + \frac{1}{\gamma}\int_0^T (\kappa_t^T + \delta_t^T) dZ_t^\delta \geq 0,$$

or, equivalently,

$$\int_0^T (\kappa_t^T + \delta_t^T) dZ_t^\delta \geq -\gamma \hat{W}_0 B_T,$$

$$\int_t^T (\kappa_s^T + \delta_s^T) dZ_s^\delta \geq -\left(\gamma \hat{W}_0 B_T + \int_0^t (\kappa_s^T + \delta_s^T) dZ_s^\delta\right). \quad (15)$$

With respect to Q^δ, the LHS of inequality (15) is independent of F_t and Gaussian with mean zero and variance $\int_t^T \|\kappa_s + \delta_s\|^2 ds$. Thus, we can write

$$u \geq -\frac{\gamma \hat{W}_0 B_T + \int_0^t (\kappa_s^T + \delta_s^T) dZ_s^\delta}{[\int_t^T \|\kappa_s + \delta_s\|^2 ds]^{1/2}} \equiv -d_t,$$

where u is the standard (0,1) normal. It is important to note that d_t is a function of optimal unconstrained wealth at time t, as will be shown later on.

Therefore, $E^{Q^\delta}[1_A | F_t) = N(d_t)$.

Finally, as shown in the appendix, the last conditional expectation in equation (14) is written

$$E^{Q^\delta}\left[\left(\frac{1}{\gamma}\int_t^T (\kappa_s^T + \delta_s^T) dZ_s^\delta\right) 1_A | F_t\right] = n(d_t)\left[\int_t^T \|\kappa_s + \delta_s\|^2 ds\right]^{1/2},$$

where $n(\cdot)$ is the normal density function. Thus

$$W_t^* = \hat{W}_t N(d_t) + \frac{1}{\gamma}\frac{B_t}{B_T} n(d_t)\left[\int_t^T \|\kappa_s + \delta_s\|^2 ds\right]^{1/2}. \tag{16}$$

It remains to derive the dynamics of optimal "constrained" wealth at any time t. Applying Itô's lemma to equation (16) and substituting $\hat{W}_t$ by its expression in equation (12), we get

$$dW_t^* = g(\cdot)dt + N(d_t)\frac{1}{\gamma}\frac{B_t}{B_T}(\kappa_t^T + \delta_t^T) dZ_t, \tag{17}$$

where $g(\cdot)$ is the (rather complicated) drift of the wealth process whose exact value is of no interest here.

In order to derive the vector θ^* of optimal positions in the futures contracts, we now identify the dynamics of wealth given by equation (7) with that given by equation (17). As seen previously in the logarithmic utility case, it is sufficient to identify the diffusion coefficients.[13] Thus,

$$\Pi^T I_{S_t}\sigma_t + \theta_t^T I_{F_t} v_t = N(d_t)\frac{1}{\gamma}\frac{B_t}{B_T}(\kappa_t^T + \delta_t^T). \tag{18}$$

Some comments are in order. The first term in the LHS comes from the fixed position in the primitive assets, while the second can be adjusted through futures trading. The RHS represents the CARA investor's optimal demand in a fictitiously completed market where the market price of risk, $\kappa + \delta$, is deterministic. Although the investor's utility function differs from the myopic logarithmic case, the optimal hedging demands (18) do not include a third component à la Breeden-Merton—that is, an informationally based hedging component associated with shifts in the opportunity set. But this of course is because we have assumed deterministic diffusion price coefficients. Finally, the demand is corrected by a term reflecting the binding nonnegativity constraint on wealth.

Again, we separate equation (18) into its two orthogonal components in Range (v_t) and Ker (v_t) to get, respectively,

$$\Pi^\tau I_{S_t}\sigma_t v_t^\tau (v_t v_t^\tau)^{-1} v_t + \theta_t^\tau I_{F_t} v_t = N(d_t)\frac{1}{\gamma}\frac{B_t}{B_T}\kappa_t^\tau \tag{19}$$

$$\Pi^\tau I_{S_t}\sigma_t (I_L - v_t^\tau (v_t v_t^\tau)^{-1} v_t) = N(d_t)\frac{1}{\gamma}\frac{B_t}{B_T}\delta_t^\tau. \tag{20}$$

In equation (20), we assume that there exists a price for orthogonal risks that makes the optimal demand on the RHS equal to the fixed position on the LHS. Denoting δ_t^* this process, we turn to equation (19) in which it is readily seen that the investor can fill the gap between the exogenous and the optimal demands of traded risks by taking the position on futures contracts given by

$$\theta_t^\tau I_{F_t} = -\Pi^\tau I_{S_t}\sigma_t v_t^\tau (v_t v_t^\tau)^{-1} + \frac{1}{\gamma}\frac{B_t}{B_T} N(d_t)(\eta_t - r_t 1)^\tau (v_t v_t^\tau)^{-1}. \tag{21}$$

Equation (21) deserves the following remarks. The first term on the RHS is the "pure hedge" component of the strategy. If the markets were complete and the interest rate deterministic, we would have $I_{F_t} v_t = (B_T/B_t) I_{S_t}\sigma_t$ by arbitrage, so that this component of θ_t^τ would be equal to $(B_t/B_T)\Pi^\tau$, the traditional result except for the discount factor due to the continuous marking-to-market of the futures position. Here, futures and cash instruments are not perfectly correlated so that the absolute value of this "pure hedge" component is smaller than would be the case in the complete market case. This result is intuitive since the less futures contracts are correlated with the exposed, nontraded, primitive assets, the less attractive is hedging.

The second term on the RHS, the "speculative" component, reflects the usual mean-variance tradeoff. It differs from the one encountered in the literature for two reasons. The first one is the same as above—namely, the imperfect correlation between cash and future,[14] which also lowers in absolute terms this component of the optimal hedging demands. The second reason is that imposing a nonnegativity constraint on wealth introduces the extra term $N(d_t)$. This term is reminiscent of the Black-Scholes formula. This does not come as a surprise since, as we have seen previously, the investor implicitly must buy a protective European put written on his unconstrained wealth. However, it is important to recall that the unconstrained terminal wealth $\widehat{W}_T$, which may be negative and which is the underlying asset of the put, is not lognormally but normally distributed under both measures P and Q^δ. Also, since $N(d_t) < 1$, the speculative demand for futures is smaller, in absolute terms, than the one that would obtain without the nonnegativity constraint on wealth. This is because being long in the put is equivalent to being short in the underlying asset. The intuition behind this result is that the investor must impose limits on his trading of futures contracts without which his wealth could become negative.

Equation (21) also reveals a major difficulty ignored in the literature: even in the case of geometric Brownian motions and perfect correlation between futures and spot assets, it is not possible to derive "closed-form" solutions for the optimal hedging demands, in the sense that θ depends on d_t and d_t depends, not on effective wealth W_t but on "unconstrained" wealth $\hat{W}_t$. It is not indeed possible to recover W_t since the relation between $\hat{W}_t$ and W_t is not invertible. Thus, we have shown that only for those utility functions that exhibit infinite marginal utility at zero wealth (such as the logarithmic), can we get closed-form solutions in the aforementioned sense. Note however that equation (21) is not worthless. Indeed it is possible to solve with numerical methods for $\hat{W}_0$, then $\hat{W}_t$ and then θ_t^*.

Finally, it would be wrong to read (21) as if the vector δ_t^* played no role. Although the optimal solution looks like its 'counterpart" in complete markets, it is misleading since s_t^* *is embedded implicitly in* d_t. This makes the numerical procedure mentioned above even more complicated.

6. Concluding remarks

This article has shown how to adapt the methodology of He and Pearson [1991] and Karatzas et al. [1991] relative to the optimal portfolio choice problem under incomplete markets to solve an optimal hedging problem. We have examined the two cases where (1) the investor's utility function is logarithmic and prices of primitive assets and futures contracts follow diffusion processes and (2) the investor's utility function is CARA and the market prices obey less general processes with deterministic parameters.

When the investor's utility function is logarithmic, his optimal hedging demands for futures contracts have been derived in closed form and shown to resemble formally those that would obtain under complete markets. In particular, they do not exhibit Merton-Breeden components due to the investor's myopia. The two differences are that futures and spot instruments are not perfectly correlated and that incompleteness leads to a second best optimum.

When utility functions exhibit CARA, the optimal hedging demands for futures cannot be derived in closed form, even when we force the market prices of risk to be deterministic, but otherwise the same conclusions hold as in the logarithmic case. The absence of Merton-Breeden components is in this case due to the assumed absence of shifts in the investment opportunity set. There are two reasons why a closed-form solution cannot be attained. First, the nonnegativity constraint on wealth is here binding, which prevents us from deriving the hedging demands explicitly in terms of the investor's actual wealth, making numerical procedures compulsory. Second, the incompleteness of markets creates complications.

As far as equilibrium is concerned, aggregation through investors in our setting, and thus the derivation of open interest in each futures contract for investors constrained on their holdings of spot assets, is clearly untractable, under the CARA assumption. It was already the case under complete markets, as shown by Lioui and Poncet [1996], due to the presence of nonnegativity constraints on wealth.

Potential extensions of our framework would include the following:

- Primitive securities could be allowed to pay out dividends. This extension is easy if the latter are continuous but problematic if they are discrete.
- Maturity dates T_j of the futures contracts ($j = 1, \ldots, N$) could differ across j and not coincide with the investor's horizon T. Note that this would be another way to introduce imperfect correlation between primitive assets and futures contracts, even though each

futures were written on a particular primitive asset, because of basis risk. When $T < T_j$, basis risk emerges directly; when $T > T_j$, the need for a "rolling hedge" will arise up to the last contract for which $T < T_j$. All these aspects are clearly relevant to current practice, but would aggravate the mathematical burden without necessarily leading to a better economic understanding of the main issue at stake under incomplete markets.

- The market price of risk for utility functions other than the logarithmic could be made stochastic. This is a serious problem, however, the solution to which would involve at least a third term à la Merton-Breeden in the optimal futures demands.
- Utility functions more general than the two examined here could be assumed. This problem seems tractable and is under progress.

Appendix

We compute the term $L \equiv E^{Q^\delta}[(\int_t^T(\kappa_s^T + \delta_s^T)dZ_s^\delta)1_A | F_t]$ that appears in equation (14) in the text.

Recall that $\int_t^T(\kappa_s^T + \delta_s^T)dZ_s^\delta$ is independent of F_t and Gaussian with respect to the measure Q^δ of mean zero and variance $\int_t^T \| \kappa_s + \delta_s \|^2 ds$. Then

$$L = E^{Q^\delta}\left[\left(\int_t^T (\kappa_s^T + \delta_s^T)dZ_s^\delta\right) 1_A\right]$$

$$= E^{Q^\delta}[u 1_A] \cdot \left[\int_t^T \| \kappa_s + \delta_s \|^2 ds\right]^{1/2} \quad \text{(from the definition of } u \text{ and } d_t \text{ in the text)}$$

$$= \left[\int_t^T \| \kappa_s + \delta_s \|^2 ds\right]^{1/2} \int_{-d_t}^{\infty} \frac{1}{\sqrt{2\pi}} u e^{-(u^2/2)} du.$$

Let $y \equiv u^2/2$. Then

$$L = \left[\int_t^T \| \kappa_s + \delta_s \|^2 ds\right]^{1/2} \int_{d_t^2/2}^{\infty} \frac{1}{\sqrt{2\pi}} e^{-y} dy = \left[\int_t^T \| \kappa_s + \delta_s \|^2 ds\right]^{1/2} n(d_t).$$

Acknowledgments

We thank Steinar Ekern and an anonymous referee for their useful and competent remarks, which have improved the overall quality of this article. We also have benefited from discussions with our colleagues of the Finance Department at ESSEC and the participants of the European Group of Risk and Insurance Economists, Geneva, September 18–20, 1995. All remaining errors are ours.

Notes

1. The condition $N < L$ is necessary but not sufficient to have true market incompleteness. If the parameters of the price process are functions only of the current prices of the traded securities (or even simpler functions), we are back to an essentially complete market, as in the Black-Scholes environment where geometric Brownian motions are assumed. This limited the scope of Pagès' work.

2. See, however, note (1).
3. We briefly discuss in Section 6 the implications of relaxing this coincidence assumption.
4. $N < M$ is not required, provided that both N and M are strictly smaller than L and primitive assets and futures are not perfectly correlated. Our analysis covers also this $N > M$ case.
5. If to each and every primitive asset were associated a futures contract perfectly correlated to it, then (markets would be "quasi-complete" and) absence of arbitrage opportunities would insure that, as in Lioui and Poncet [1996]: $\eta_t = \mu_t - r_t 1$ and $v_t = \sigma_t$.
6. For technical conditions satisfied by these strategies, see Cox and Huang [1989].
7. From the seminal work of Harrison and Kreps [1979] and Harrison and Pliska [1981], extended by Geman, El Karoui, and Rochet [1995] and Duffie [1992], there exist such martingale measures, given any appropriate numéraire. Also, it is unique when markets are complete.
8. Although the constrained investor can obviously observe the information conveyed by μ_t and σ_t present in equation (1), any market price of risk process that would be defined by μ_t and σ_t would be irrelevant to him since he cannot freely trade on the primitive assets.
9. In Cox and Huang [1989], the tangent portfolio is just the numéraire portfolio H.
10. We implicitly assume that equation (10) admits a solution in the space of orthogonal risk premia processes.
11. Recalling the definition of κ_t, the dimension of the term multiplying W_t^* is expected risk premia over variance of such premia.
12. Had we assumed more generally κ_t and δ_t to be stochastic, we would have ended up with a partial differential equation for the optimal wealth (such as Cox and Huang [1989] in their general framework) without being able to obtain an explicit solution.
13. Note that direct identification of the drift terms is almost untractable due, in particular, to the presence of X_t in the drift term of equation (7). This explains why we did not have to be explicit about $g(\cdot)$ in equation (17).
14. This feature is present also in Duffie and Jackson [1990].

References

ADLER, M., and DETEMPLE, J. [1988a]: "Hedging with Futures in an Intertemporal Portfolio Context," *The Journal of Futures Markets*, 8, 249–269.

ADLER, M., and DETEMPLE, J. [1988b]: "On the Optimal Hedge of a Non-Traded Cash Position," *Journal of Finance*, 43, 143–153.

ANDERSON, R.W., and DANTHINE, J.P. [1983]: "The Time Pattern of Hedging and the Volatility of Futures Prices," *Review of Economic Studies*, 50, 249–266.

BAJEUX-BESNAINOU, I., and PORTAIT, R. [1994]: "The Numéraire Portfolio: A New Approach to Financial Theory," CERESSEC Working Paper.

BREEDEN, D.T. [1979]: "An Intertemporal Asset Pricing Model with Stochastic Consumption and Investment Opportunities," *Journal of Financial Economics*, 7, 266–296.

BREEDEN, D.T. [1984]: "Futures Markets and Commodity Options: Hedging and Optimality in Incomplete Markets," *Journal of Economic Theory*, 32, 275–300.

COX, J., and HUANG, C.F. [1989]: "Optimal Consumption and Portfolio Policies When Asset Prices Follow a Diffusion Process," *Journal of Economic Theory*, 49, 33–83.

COX, J., and HUANG, C.F. [1991]: "A Variational Problem Arising in Financial Economics," *Journal of Mathematical Economics*, 20, 465–487.

DUFFIE, D. [1992]: *Dynamic Asset Pricing Theory*, Princeton, N.J.: Princeton University Press.

DUFFIE, D., and JACKSON, M.O. [1990]: "Optimal Hedging and Equilibrium in a Dynamic Futures Market," *Journal of Economic Dynamics and Control*, 14, 21–33.

FÖLLMER, H., and SCHWEIZER, M. [1991]: "Hedging of Contingent Claims Under Incomplete Information," in *Applied Stochastic Analysis*, Davis, M. and Elliott, R. (Eds.), Gordon and Breach, London/New York, 389–414.

GEMAN, H., EL KAROUI, N., and ROCHET, J.C. [1995]: "Changes of Numéraire, Changes of Probability Measure and Option Pricing," *Journal of Applied Probability*, 32, 443–458.

GOURIEROUX, C., LAURENT, J.P., and PHAM, H. [1995]: "Quadratic Hedging and Numéraire," CREST Working Paper.

HARRISON, M., and KREPS, D. [1979]: "Martingales and Multiperiod Securities Markets," *Journal of Economic Theory*, 20, 381–408.

HARRISON, M., and PLISKA, S. [1981]: "Martingales and Stochastic Integrals in the Theory of Continuous Trading," *Stochastic Processes and Their Applications*, 11, 215–260.

HE, H., and PEARSON, N. [1991]: "Consumption and Portfolio Policies with Incomplete Markets and Short Sale Constraints," *Journal of Economic Theory*, 54, 259–305.

HO, T.S.Y. [1984]: "Intertemporal Commodity Futures Hedging and the Production Decision," *Journal of Finance*, 39, 351–375.

KARATZAS, I., LEHOCZKY, J., and SHREVE, S. [1987]: "Optimal Portfolio and Consumption Decisions for a Small Investor on a Finite Horizon," *SIAM Journal of Control and Optimization*, 25, 1557–1586.

KARATZAS, I., LEHOCZKY, J., SHREVE, S., and XU, G.-L. [1991]: "Martingale and Duality Methods for Utility Maximization in an Incomplete Market," *SIAM Journal of Control and Optimization*, 29, 702–730.

LIOUI, A., and PONCET, P. [1996]: "Optimal Hedging in a Dynamic Futures Market with a Nonnegativity Constraint on Wealth," *Journal of Economic Dynamics and Control*, 20, 1101–1113.

LONG, J. [1990]: "The Numéraire Portfolio," *Journal of Financial Economics*, 26, 29–69.

MERTON, R.C. [1971]: "Optimum Consumption and Portfolio Rules in a Continuous Time Model," *Journal of Economic Theory*, 3, 373–413.

MERTON, R.C. [1973]: "An Intertemporal Capital Asset Pricing Model," *Econometrica*, 41, 864–888.

PAGÈS, H. [1987]: "Optimal Consumption and Portfolio Policies When Markets are Incomplete," MIT mimeo, Cambridge, MA.

PONCET, P., and PORTAIT, R. [1993]: "Investment and Hedging Under a Stochastic Yield Curve: A Two-State-Variable, Multi-Factor Model," *European Economic Review*, 37, 1127–1147.

RUBINSTEIN, M. [1976]: "The Strong Case for the Generalized Logarithmic Utility Model as the Premier Model of Financial Markets," *Journal of Finance*, 21, 551–571.

STULZ, R.M. [1984]: "Optimal Hedging Policies," *Journal of Financial and Quantitative Analysis*, 19, 127–140.

SVENSSON, L.E.O., and WERNER, I.M. [1993]: "Nontraded Assets in Incomplete Markets: Pricing and Portfolio Choice," *European Economic Review*, 37, 1149–1168.

The Geneva Papers on Risk and Insurance Theory, 21: 123–141 (1996)

Fairly Priced Deposit Insurance, Incentive Compatible Regulations, and Bank Asset Choices

SUK HEUN YOON
Korea Institute of Finance, Coal Central Building, Fifth Floor, 80-6 Soosong-dong, Chongro-ku, Seoul 110-140, Korea

SUMON C. MAZUMDAR
Faculty of Management, McGill University, 1001 Sherbrooke St. West, Montreal, Quebec, Canada, H3A 1G5

Abstract

This article provides incentive compatible regulations that support fairly priced deposit insurance in a competitive banking industry. If informational asymmetry exists between the regulator and banks regarding loan quality, but the regulator can observe actual loan rates charged, then imposing a capital requirement schedule that leads market loan rates to decrease in loan quality is shown to be incentive compatible. Competition in the loan market induces banks to be indifferent to all loans that satisfy a minimum acceptable quality and reject all riskier loans. The regulator could reduce the banking industry's riskiness by imposing stricter capital requirements that increase this minimum quality.

Key words: deposit insurance, incentive compatibility.

1. Introduction

The dramatic escalation of U.S. bank failures, from eight to over 200 annually over the past two decades, and the recent collapse of the thrift industry resulting in an estimated deficit of $250 billion for the Federal Savings and Loans Insurance Corporation (FSLIC), demonstrate the urgent need for a reevaluation of the regulatory policies governing depository financial institutions (see Jaffee [1989, Table 1]). It is therefore not surprising that this issue has, of late, attracted considerable attention from both scholars and policy makers.

Bank regulatory reforms either in place in the United States, or suggested to date, may be broadly classified into two categories—(1) imposing risk-based capital requirements and (2) making the deposit insurance premium structure risk-adjusted.[1] These proposals are intended to reduce banks' incentives for excessive risk-taking that are believed to exist under the current flat-rate insurance premium, as well as reduce the insurer's uncompensated liabilities. However, a survey of the burgeoning literature in this area indicates that there is, as yet, little consensus regarding regulatory policies that would effectively attain the intended objectives.

The implementation of risk-adjusted regulatory policies is a difficult task since the risk of bank assets (loans) and bank managerial efforts are usually unobservable to regulator. We focus on the regulator's concern about the quality of a loan a bank chooses when it has less information than the bank regarding the loan's quality. That is, we examine incentive compatible regulatory policies that would induce the bank to only choose, and truthfully report, loans of certain quality to start with, rather than to improve its managerial efforts afterwards.[2]

Our article thus addresses the same question analyzed by Chan, Greenbaum, and Thakor [1992] insofar as we too are interested in establishing incentive compatible regulatory policies that would allow the deposit insurance system to break even. However, our conclusion is critically different from theirs. Chan, Greenbaum, and Tahkor [1992] argue that fairly priced deposit insurance cannot, in general, be consistent with incentive compatible regulations in a competitive banking industry.[3] In contrast, we show that if the regulator observes the loan rates that individual banks actually charge their loan customers in a competitive market, then incentive compatible regulations that guarantee fairly priced deposit insurance could exist. Such observability would force the loan rates actually charged by banks to fully signal their true qualities in the competitive loan market if and only if an intuitively appealing condition for banks' capital requirements is met.

The article has two objectives. We first examine whether or not the regulator could impose incentive compatible regulations, that are also actuarially fair, on banks operating in a competitive loan market. Second, we analyze the regulator's control of banks' asset choices through such regulations.

In the model, banks compete for loans to borrowers who are characterized by the success probability (quality) of their investment projects. This loan quality is observed by the bank but not by the regulator. Therefore, the regulator relies on a bank's loan quality report to assess the required capital and premium for full coverage deposit insurance. However, the regulator monitors the bank and observes the loan rate it actually charges its customer. In a competitive loan market, observing such loan rates should not be a difficult task for the regulator because these rates are normally well advertised (for example, car loans, mortgage loans, and prime loans).[4] The regulator's access to such information then requires the bank to be consistent between the loan quality reported and the loan rate actually charged.

We demonstrate that if and only if the competitive loan rate schedule is a *decreasing* function of loan quality, then the bank must truthfully reveal its loan quality. However, competitive loan rates are shown to critically depend on bank capital requirements. Therefore, the regulator's task is to design a capital requirement schedule that leads loan rates to be *decreasing* in loan quality. Any such regulatory scheme would guarantee incentive compatibility and also enable the fair pricing of deposit insurance.

Such incentive compatible regulations yield three important policy implications. First, the deposit insurance system could break even. This would ease the current financial strain on the FDIC's Bank Insurance Fund. Second, competition in the loan market would drive down the market loan rates for all loans, so that in equilibrium individual banks should earn at most zero net present value from any loan. That is, banks would not necessarily favor a riskier loan project. Third, our model lends support to the conventional argument that stricter capital requirements would discipline banks in their asset choices. Banks would reject loans that do not satisfy a minimum acceptable quality below which the loan's net present value to bank equityholders becomes negative. In this situation, an upward shift in the capital requirement schedule would increase this minimum acceptable quality. As a result, the overall riskiness of a bank's loan portfolio, as well as that of the whole banking industry, would be lower.

In Section 2, we describe the model. In Section 3, we discuss the bank's financing decision and establish its optimal capital structure. In Section 4, we discuss the existence and characterization of incentive compatible regulations that guarantee fairly priced deposit insurance. In Section 5, we analyze the implications of such incentive compatible policies on the bank's asset choice and discuss the role of regulations in improving the "safety and soundness" of the banking industry. Section 6 concludes. All proofs are contained in the appendix.

2. The model

Consider a capital market with many risk-neutral firms and banks. Firms have investment projects specified below and are assumed to finance their projects completely through bank loans. Banks take deposits and make loans to finance firms' projects. This capital market is competitive in that a bank, in equilibrium, earns zero net present value from lending.[5] In the market, one plus the risk-free interest rate is given by r. We assume $r \geq 1$.

We consider a one period economy with two event dates, $t = 0$ and $t = 1$. A firm is identified by the success probability (quality) $\theta \in [0, 1]$ of its investment project. This investment project requires an initial outlay of \$1 and will produce α if successful and 0 if unsuccessful, where α is a finite constant with $\alpha \geq$ r. At $t = 0$, a bank can observe the success probability of a firm's project, θ, while the actual success or failure of the project is observed at $t = 1$.

Banks operating in this capital market are subject to deposit-linked government regulations and pay corporate taxes. We let $T \in (0, 1)$ be the corporate tax rate common to all banks. Investors in this capital market do not pay personal taxes.[6] The success probability θ of a firm's project for which a bank makes a loan is assumed to be unobservable by any third party including the regulator. In this situation, the regulator must rely on its observation of the contractual loan rate the bank actually charges its customer, and the bank's loan quality report, $\hat{\theta}$, to assess the bank's regulatory requirements.[7] That is, the regulator charges an insurance premium of $\pi(\hat{\theta})$ per dollar deposit and requires the bank's equityholders to put up a minimum capital of $e(\hat{\theta})$ per dollar loan. The deposit insurance is assumed to be provided in full. Therefore, the interest rate on deposits is the risk-free rate.

Two clarificatory comments regarding our model are in order.[8] First, we assume that the bank's loan portfolio consists of a single loan and its price is fully characterized by the interest rate charged. Such a framework is commonly adopted to tractably analyze the role of bank regulations under asymmetric information (see Giammarino, Lewis, and Sappington [1992], Chan, Greenbaum, and Thakor [1992], Besanko and Kanatas [1990], among others.) However, it should be noted that the loan's "price" often consists of the explicit loan rate as well as additional terms such as collateral and compensating balances. Although some of these embedded options may not be explicitly defined in loan documents, our results would be qualitatively unaffected if these terms continue to be dictated by competitive market forces and the regulator's audit was assumed to fully reveal these additional terms attached to any loan contract. In short, it is the relative ease with which the regulator can observe the competitive loan pricing terms compared to the loan's actual quality that is critical to our analysis.

Second, the model assumes that collusive side-payments between the bank and the borrower, to take advantage of any regulatory subsidy, are not possible. Such a framework is tenable only if the regulator can observe and prevent such collusion without bearing substantial supervisory costs. Mazumdar and Yoon [forthcoming] demonstrate that if such collusive side-payments cannot be prevented, then incentive compatible regulation design, together with fairly priced deposit insurance, will be impossible in an asymmetric information scenario similar to ours.[9] This raises an important policy issue. Any consideration of incentive compatible regulatory design must first examine the relative size of the potential regulatory subsidy saving compared to the supervision costs necessary to prevent bank-borrower collusion. We do not formally examine such a tradeoff in this paper.

In our framework, the following sequence of activities is viewed to occur at $t = 0$. First, a firm with a success probability θ applies for a loan from a bank in the competitive banking industry. The bank evaluates this loan application by taking into account its subsequent choices of θ and the financing mix described below. The bank then decides whether to grant the loan or not and offers a contractual loan rate, R, if it chooses to grant it. Second, when the loan contract is finalized, the loan rate R actually charged by the bank is observed by the regulator. At that time, the bank reports $\hat{\theta}$ to the regulator who then imposes $e(\hat{\theta})$ and $\pi(\hat{\theta})$ on the bank. Third, given θ, $\hat{\theta}$, R, $e(\hat{\theta})$ and $\pi(\hat{\theta})$, the bank chooses the optimal levels of deposit, D, and paid-in equity, E, with which to finance the contracted loan.

At $t = 1$, if the project is successful, the firm successfully pays off R to the bank and retains the net earning $\alpha - R$. The bank then uses its proceeds to pay off its deposit liabilities, rD, its corporate tax bill, $(R - (r - 1)D)T$, and a liquidating dividend to its shareholders. Clearly, the deposit insurer has no liability in this successful state.

But if the project fails, then the firm defaults completely on the loan and the bank declares bankruptcy. The bank depositors are then paid off rD by the deposit insurance agency. In what follows, we develop our model in an order reverse to the sequence just described since we analyze the bank's decision problem recursively.

3. Bank's financing decision

Consider a bank that has already contracted a loan of quality θ, offered a loan rate R, reported $\hat{\theta}$ to the regulator, and also has been assigned $e(\hat{\theta})$ and $\pi(\hat{\theta})$ by the regulator. The bank's budget constraint is

$$1 + \pi(\hat{\theta})D = D + E, \quad \text{and} \tag{1}$$

$$E \geq e(\hat{\theta}), \tag{2}$$

where $e(\hat{\theta})$ is the bank's capital requirement.

In equation (1), the demand for deposits by the public is assumed to be infinitely elastic at the given r, and the bank is allowed to pay its deposit insurance premium $\pi(\hat{\theta})D$ using deposits as long as $E \geq e(\hat{\theta})$.[10]

The bank maximizes the net present value of its current equityholders, $NPV^E(\theta\ \hat{\theta})$, by choosing its financial structure, or simply E, given θ, R, $\hat{\theta}$, $e(\hat{\theta})$ and $\pi(\hat{\theta})$. That is, the bank's problem is

$$\max_E NPV^E(\theta, \hat{\theta}) \equiv \frac{\theta[R(1 - T) - rD + (r - 1)DT]}{r} - E, \tag{3}$$

subject to (1) and (2).

In (3), the first term in the RHS is the true value to the bank's currert equityholders, which is not observed by any outsiders (including both the external bank equity market and the regulator). The cost of equity capital is r in this risk-neutral economy without personal taxes. The second term is the current equityholders' contribution to the loan.[11]

Let us define $s(\theta, \pi(\hat{\theta}))$ to be the net value of deposit insurance subsidy per dollar deposit given θ and $\hat{\theta}$. It is clear that the total net value of deposit insurance subsidy $s(\theta, \pi(\hat{\theta}))D$ equals

$$\frac{(1 - \theta)rD}{r} - \pi(\hat{\theta})D = (1 - \theta - \pi(\hat{\theta}))D. \tag{4}$$

The first term in the LHS of (4) is the present value of the insurer's contingent liability given the true failure probability. The second term is the total insurance premium the bank pays. The deposit insurance premium schedule based on $\hat{\theta}$ is said to be risk-based if

$$\pi(\hat{\theta}) = 1 - \hat{\theta} \quad \forall \ \hat{\theta} \in [0, 1]. \tag{5}$$

Note that if the bank reports truthfully—that is, $\hat{\theta} = \theta$—then such a risk-based deposit insurance premium would be actuarilly fair—that is, $s(\theta, \pi(\hat{\theta})) = 0$.

Lemma 1. *Assume the regulator adopts schedule (5). Given θ, $\hat{\theta}$, R, $e(\hat{\theta})$, and $\pi(\hat{\theta})$, the bank's optimal financing mix becomes*

$$E(\hat{\theta}) = e(\hat{\theta}) \ \text{ and } \ D(\hat{\theta}) = \frac{1 - e(\hat{\theta})}{\hat{\theta}} \ \text{ if } \ \hat{\theta} > \theta^*; \tag{6}$$

$$E(\hat{\theta}) = 1 \ \text{ and } \ D(\hat{\theta}) = 0 \ \text{ if } \ \hat{\theta} < \theta^*; \ \text{ and} \tag{7}$$

the bank's financing mix becomes irrelevant if $\hat{\theta} = \theta^*$; where $\theta^* \equiv (1 - (r - 1)T/r)\theta < \theta$.

When the bank uses deposits to finance its loan, it benefits from two different types of subsidies: (1) a tax-related subsidy arising from the tax deductibility of interest expenses and (2) a deposit insurance related subsidy arising from a potential mispricing of the deposit insurance premium due to a misrepresentation of the loan quality.[12] Lemma 1 points out that what the bank has reported will in turn govern its financing behavior. If $\hat{\theta} > \theta^*$, then the combined value of the two subsidies is positive. Hence, the bank would increase its deposits up to the maximum level of $[1 - e(\hat{\theta})]/\hat{\theta}$, in order to take advantage of these subsidies. If $\hat{\theta} < \theta^*$, then the combined value of the subsidies is negative, and hence the bank would finance its loan entirely through equity capital. Finally, if $\hat{\theta} = \theta^*$, then the combined value is zero, and hence the bank would be indifferent to the two financing methods.

We define $R(\theta)$ as the risk-adjusted market loan rate that a bank, if it truthfully reported its loan quality θ, must charge in order to break even. Recall that R denotes the *actual* loan rate the bank charges for a loan of quality θ. Competition in the loan market dictates that $R \leq R(\theta)$. The bank would otherwise lose the loan customer.

Proposition 1. *A bank making a loan of quality θ will always report $\hat{\theta} > \theta^*$ and employ the financing mix given by* (6).

The intuition behind this proposition is straightforward. If the bank reports $\hat{\theta} < \theta^*$, then the combined value of the two subsidies is negative, and thus it cannot offer competitive loan rates without incurring a loss. Note that the bank cannot increase its contractual loan rate R beyond the market loan rate $R(\theta)$ in an attempt to recover this loss. Therefore, in order to offer competitive loan rates, the bank must take advantage of the two subsidies associated with deposits. Accordingly, the bank would always report $\theta > \theta^*$. Lemma 1 then yields that the bank would always employ the maximum deposits allowed by $e(\hat{\theta})$. This proposition thus suggests that the bank in this economy would effectively function as a depository financial institution.

To conclude this section, suppose the overall regulatory package is incentive compatible, so that the bank truthfully reports θ to the regulator. In this case, the deposit insurance subsidy will disappear, but the bank will still increase its deposits to the maximum allowed level of $[1 - e(\theta)]/\theta$ to take full advantage of the interest tax shield.

4. Incentive compatible regulations

The purpose of deposit-linked regulations for banks is viewed to be twofold. The first is to ensure that full-coverage deposit insurance is fairly priced. The second is to control banks' lending behavior in terms of their choices of risky assets (loans). In this section, we analyze incentive compatible regulations that allow the regulator to charge fair insurance premiums.[13] We will discuss the regulator's control of banks' asset choices in the next section.

We now define $R(\theta, \hat{\theta})$ to be the loan rate that a bank applies to a loan application of quality θ, when the bank reports $\hat{\theta}$ to the regulator. Note that $R(\hat{\theta}, \hat{\theta}) \equiv R(\hat{\theta})$ for all $\hat{\theta}$, where $R(\hat{\theta})$ is the risk-adjusted market loan rate for a loan of quality $\hat{\theta}$ as defined in the previous section.[14] Upon receiving such a report, the regulator infers the following: had the bank been truthful, then it would have only offered $R(\hat{\theta})$ to its customer because a truthful bank would otherwise either have lost the customer (if $R(\theta, \hat{\theta}) > R(\hat{\theta})$) or have incurred a loss (if $R(\theta, \hat{\theta}) < R(\hat{\theta})$). The regulator's observability of the actual loan rates then forces such an untruthful bank, with $\hat{\theta} \neq \theta$, to offer

$$R(\theta, \hat{\theta}) = R(\hat{\theta}) \tag{8}$$

so that the actual loan rate charged conforms to its reported value $\hat{\theta}$. We write the net present value of the current equityholders of such an untruthful bank as,

$$NPV^E(\theta, \hat{\theta}) = \frac{\theta[R(\hat{\theta})(1 - T) - (r - (r - 1)T)D(\hat{\theta})]}{r} - E(\hat{\theta}). \tag{9}$$

In (9), the bank offers the loan rate $R(\hat{\theta})$ and employs deposits $D(\hat{\theta})$ and equity $E(\hat{\theta})$ to be consistent with $\hat{\theta}$ since all these three variables are observed by the regulator.

On the other hand, we write the net present value of the current equityholders of a truthful bank as

$$NPV^E(\theta) = \frac{\theta[R(\theta)(1 - T) - (r - (r - 1)T)D(\theta)]}{r} - E(\theta). \tag{10}$$

Note in both (9) and (10) that the bank's net lending cost is its equity contribution, which also affects its net payoffs at $t = 1$. Substituting the capital structure of such a truthful bank, from (6), into $D(\theta)$ and $E(\theta)$ in (10) and solving $NPV^E(\theta) = 0$, we can get the risk-adjusted market loan rate schedule as

$$R(\theta) = \frac{e(\theta)r + (r - (r - 1)T))(1 - e(\theta))}{(1 - T)\theta} \quad \forall \quad \theta \in [0, 1]. \tag{11}$$

Two aspects of this schedule are noteworthy. First, the loan rate schedule critically depends on the bank's capital requirement schedule e chosen by the regulator, which dictates the maximum deposits allowed for the bank. The bank then passes on the tax-related subsidy associated with such deposit financing to its customer in an effort to offer a competitive loan rate. Second, given that α is finite, it is possible that $R(\theta) > \alpha$ for some θ. For now, we define Θ to be the set of θ such that $R(\theta) \leq \alpha$. For the remainder of this section we focus only on $R(\theta)$ for $\theta \in \Theta$. We postpone discussion of this set Θ to the next section.

The regulator's problem may now be stated as follows. The regulator must observe $R(\theta, \hat{\theta})$ and choose a pair of schedules $\{e(\hat{\theta}), \pi(\hat{\theta})\}$ for all $\hat{\theta} \in \Theta$ such that

$$\pi(\hat{\theta}) = 1 - \hat{\theta}; \tag{12}$$

$$D(\hat{\theta}) = \frac{1 - e(\hat{\theta})}{\hat{\theta}} \quad \text{and} \quad E(\hat{\theta}) = e(\hat{\theta}); \quad \text{and} \tag{13}$$

$$NPV^E(\theta) \geq NPV^E(\theta, \hat{\theta}) \quad \forall \quad \theta \quad \text{and} \quad \hat{\theta}. \tag{14}$$

Equation (12) gives a risk-based deposit insurance schedule. Equation (13) provides the solutions to the bank's financing decision, as obtained in Proposition 1. Equation (14) states the incentive compatibility condition that would induce the bank to voluntarily report the true value of θ to the regulator.

Lemma 2. *Assume the regulator adopts schedule $\pi(\theta) = 1 - \theta$. The market loan rate schedule $R(\theta)$ in* (11) *becomes decreasing (constant, increasing) in θ for all $\theta \in \Theta$ if and only if*

$$e'(\theta) < (=, >)Q(\theta) \equiv \frac{r - (1 - e(\theta))(r - 1)T}{\theta(r - 1)T} \quad \forall \quad \theta \in \Theta, \tag{15}$$

where $e'(\theta) \equiv \partial e(\theta)/\partial\theta$ and $Q(\theta) > 0$.

$Q(\theta)$ in (15) represents the marginal increase in the capital requirement required to induce the market loan rate to be constant in θ. To understand this term better, let us consider the case when $e'(\theta) = Q(\theta) > 0$. In this case, for a loan of a higher quality, the capital requirement becomes greater and the insurance premium becomes smaller, both of which contribute to the reduction in the deposits used. However, the bank equity's net present value remains unchanged for the higher quality. This is because the decrease in the value due to the greater capital requirement is just offset by its increase due to the decrease in the bank's contingent liability value (resulting from reduced deposits). Thus, the market loan rate should remain unchanged for the higher quality loan. For a higher quality loan, any incremental capital requirement which is greater than $Q(\theta)$ will force the market loan rate to be higher, and vice versa.

As will be shown shortly, a market loan rate schedule that is downward sloping throughout in θ is essential for the regulatory package $\{e, \pi\}$ to be incentive compatible. In this context, Lemma 2 makes an important contribution by demonstrating how the regulator can appropriately select a capital requirement schedule that would yield such a downward-sloping market loan rate schedule.

Proposition 2. *Assume the regulator adopts schedule (12). The regulatory requirement schedules $\{e, \pi\}$ would result in individual banks' truthfully reporting their loan qualities to the regulator if and only if the associated market loan rate schedule is decreasing throughout in θ—that is,*

$$e'(\hat{\theta}) < Q(\hat{\theta}) \quad \forall \quad \hat{\theta} \in \Theta. \tag{16}$$

When the regulator observes the loan rates applied by banks operating in the competitive loan market and the insurance premium satisfies condition (12), this proposition identifies condition (16) as necessary and sufficient for the regulatory requirement schedules $\{e, \pi\}$ to be incentive compatible and hence provide actuarially fair deposit insurance.

We refer to Figure 1(a) to discuss how a decreasing market loan rate schedule (that is, condition (16)) is *sufficient* for regulations to be incentive compatible. Consider a bank making a loan of quality θ_t. This bank cannot charge $R(\hat{\theta}_u)$ ($> R(\theta_t)$) to the customer due to competition. Since the regulator observes the loan rates actually charged, the bank is forced not to underreport $\hat{\theta}_u$. This bank also will not offer $R(\hat{\theta}_o)$($< R(\theta_t)$), which would allow it to overreport $\hat{\theta}_o > \theta_t$. Although the regulator's observability of loan rates implies

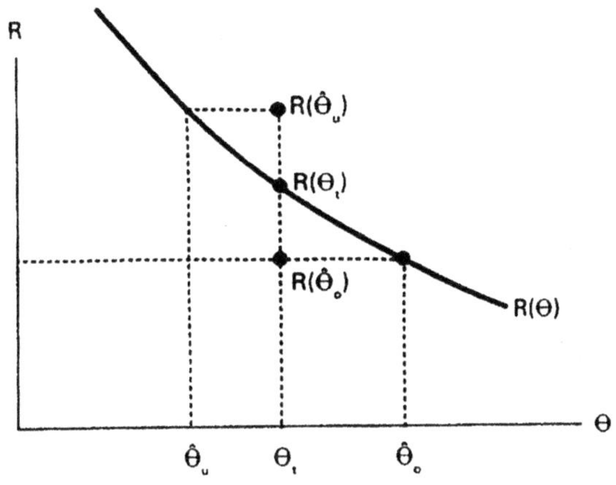

(a) A decreasing market loan rate schedule: $e'(\Theta) < Q(\Theta)$ and $\pi(\Theta) = 1 - \Theta$.

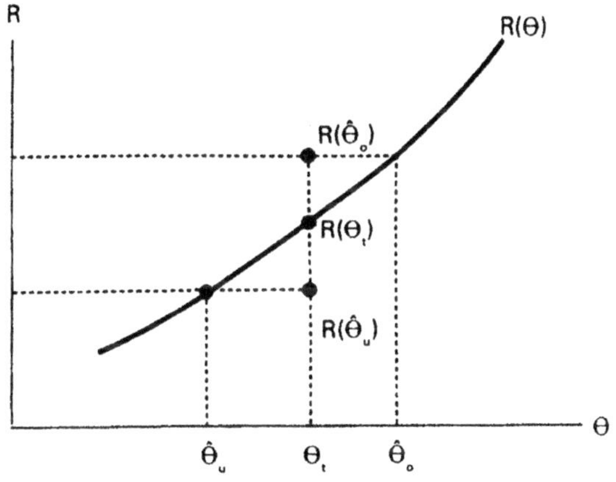

(b) An increasing market loan rate schedule: $e'(\Theta) > Q(\Theta)$ and $\pi(\Theta) = 1 - \Theta$.

Figure 1. Sufficient and necessary conditions for incentive compatibility of the loan rate schedule.

that the bank must charge $R(\hat{\theta}_o)$ if it wishes to overreport $\hat{\theta}_o$, such a loan rate would not be high enough to provide zero net present value to its current equityholders. $R(\hat{\theta}_o)$, together with $D(\hat{\theta}_o)$ and $E(\hat{\theta}_o)$, would provide zero net present value to them only if the loan quality was truly $\hat{\theta}_o$.

Condition (16) is, in fact, also *necessary* for incentive compatibility. To see this, we refer to Figure 1(b), where $R(\theta)$ is increasing. Again consider a bank making a loan of quality θ_t. This bank has an incentive to charge $R(\hat{\theta}_u)(< R(\theta_t))$ so that, by falsely underreporting $\hat{\theta}_u$ and using $D(\hat{\theta}_u)$ and $E(\hat{\theta}_u)$, the bank could make a profit. In other words, the debt-equity mix and the loan rate that would provide the bank with zero net present value for a loan of quality $\hat{\theta}_u$ would yield positive net present value when applied to the safer loan of quality θ_t. Furthermore, since $R(\hat{\theta}_u) < R(\theta_t)$, the customer would also benefit from such misreporting. However, such a bank would not charge $R(\hat{\theta}_o)$ in order to overreport $\hat{\theta}_o > \theta_t$

because if it offered $R(\hat{\theta}_o)(> R(\theta_t))$, then it would lose its customer.[15] In short, the market loan rate schedule *must* decrease throughout in loan quality for the regulations to be incentive compatible.

If $R(\theta)$ is decreasing throughout in θ, the competition in the loan market and the observability of loan rates by the regulator would force the loan rates charged by banks to fully signal their true loan qualities. Thus, the regulator could deduce the banks' true loan qualities from the loan rates they charge. On the other hand, if $R(\theta)$ is nondecreasing, then loan rates charged would not function as a signal even with competition and observability. In this sense, condition (16) can be viewed as a way through which loan rates charged by banks could be made to function as a signalling device.

Most of the literature dealing with incentive compatibility in bank regulations have suggested $e' > 0$ as a critical characterization of incentive compatible regulations (e.g., Besanko and Kanatas [1990]; Flannery [1989]. Chan, Greenbaum, and Thakor [1992] too provide a similar characterization for their incentive compatible regulations when the deposit insurance related subsidy for banks is positive.) On the contrary, we suggest $e' < Q(\theta)$; that is, incremental equity requirement for a higher quality loan need only be smaller than the positive value $Q(\theta)$. Since $e' < 0$ certainly meets this criterion, our analysis can be viewed to provide a theoretical support for recently suggested policy reforms such as the Basle Accord and the FIRREA, in which banks are required to put up more equity capital on riskier assets. In this regard, it is important to distinguish between the role played by the capital requirement schedule in earlier studies and in ours. In earlier studies a higher capital requirements for a higher loan quality serves as a direct disincentive against falsely reporting a higher loan quality. On the other hand, in our model an appropriate capital requirement schedule ensures that the market loan rate becomes a decreasing function of loan quality, so that the regulator can take advantage of its observability of competitive loan rates to rule out banks' incentive to falsely report a lower quality.[16]

5. Bank asset choices

Following the seminal work by Merton [1977], which noted the "isomorphic correspondence" between deposit insurance and a put option, many have argued that banks have an incentive to make riskier loans under the current deposit insurance system. Under such a system a bank would maximize the put option value provided by the deposit insurance by selecting the riskiest assets available. This moral hazard behavior by banks has been widely believed to be at least partly responsible for the current crisis facing depository financial institutions.[17]

In a competitive loan market where incentive compatible bank regulations are in force, we argue that such moral hazard behavior by banks will be completely resolved. To see this, let us suppose that the market loan rate schedule is downward sloping so that bank regulations become incentive compatible. Then, a bank will truthfully report its loan quality, and hence the regulator's informational problem disappears. Therefore, the only remaining issue is whether or not such a truthful bank does in fact prefer a riskier loan under such regulations. In this context, our model highlights an intuitively appealing result; that is, when the competitive loan market is in equilibrium, the bank becomes indifferent to loan quality (see Proposition 3).[18] The logic behind this result is straightforward. If a loan is riskier than others and preferred by a bank, then every other bank also prefers this loan, and hence the market loan rate applicable to the riskier loan must decrease through competition until finally, in equilibrium, the bank's equityholders earn at most zero net present value from the loan.[19] Therefore, banks no longer prefer riskier loans under incentive compatible regulations in a competitive loan market.

The only remaining problem facing the bank now is whether or not a particular loan is acceptable. If the regulatory schedules $\{e, \pi\}$ satisfy conditions (12) and (16), then competition in the loan market would yield a downward sloping market loan rate schedule. Given that α is a constant, one can then define the minimum acceptable loan quality $\underline{\theta}$ such that

$$R(\theta) \gtreqless \alpha \quad \text{for} \quad \theta \lesseqgtr \underline{\theta}. \tag{17}$$

Since the bank can only charge $Min[R(\theta), \alpha]$, the current equityholders of a bank making a loan of quality θ would earn

$$NPV^E(\theta) = (<)\ 0 \quad \text{for} \quad \theta \geq (<)\ \underline{\theta}. \tag{18}$$

For $\theta \geq \underline{\theta}$, the equityholders earn $NPV^E(\theta) = 0$ if the bank charges $R(\theta) \leq \alpha$. For $\theta < \underline{\theta}$, they earn only $NPV^E(\theta) < 0$ even if the bank charges $R(\theta) = \alpha$. Given such regulations $\{e, \pi\}$, this minimum acceptable loan quality uniquely determines the set of acceptable loan qualities mentioned earlier—that is, $\Theta = [\underline{\theta}, 1]$. The bank's lending behavior is summarized in the following proposition without proof.

Proposition 3. *Suppose the regulatory requirement schedules $\{e, \pi\}$ satisfy conditions (12) and (16). Then, a bank will accept any loan of quality $\theta \geq \underline{\theta}$, and reject all loans of quality $\theta < \underline{\theta}$.*

This proposition establishes two important results regarding banks' lending behavior. First, banks are indifferent to loans of any quality θ provided $\theta > \underline{\theta}$. As discussed above, the rationale for this result is twofold: incentive compatible regulations and price competition in the loan market. The second result is that banks will avoid extremely risky loans (that is, those of quality $\theta < \underline{\theta}$). We obtain this result because α is finite.

Proposition 4 highlights the policy issues associated with $\underline{\theta}$, which critically depends on the capital requirement schedule.

Proposition 4. *Suppose the pair $\{e, \pi\}$ satisfies conditions (12) and (16). Consider a uniform increase in the capital requirement schedule across all $\theta \in [0, 1]$ and let $\underline{\theta}^B$ and $\underline{\theta}^A$, respectively denote the minimum acceptable loan quality before and after the change. Then, $\underline{\theta}^B < \underline{\theta}^A$. As a result, a bank's quality would non-decrease and its riskiness would nonincrease.*[20]

Given that $R(\theta)$ initially equals α at $\underline{\theta}^B$, an upward shift in the capital requirement schedule would force the bank to incur a loss if it continued to lend to a project of quality $\underline{\theta}^B$. A higher θ would then be required to increase the bank equityholders' net present value up to zero. Thus, raising the capital requirement schedule would induce banks to choose, on average, safer loans. In this sense, this proposition confirms the conventional wisdom regarding the role of capital requirement.[21] Note that in our model, the quality of a bank's loan, θ, is an exogenously specified constant. Therefore, a stricter capital requirement does not affect it. However, they do lead all banks to increase their minimum *acceptable* loan quality standard. Therefore, the average riskiness of loans held by the banking industry as a whole would reduce.

To understand Propositions 3 and 4 better, we now consider a special case where $\pi(\theta) = 1 - \theta$ for all $\theta \in [0, 1]$ and $e(\theta) = k$, a constant. From Proposition 2, we know that these two schedules satisfy the incentive compatibility condition and hence yield a fairly priced deposit insurance system. We first get

$$R(\theta) = \frac{k(r-1)T + (r - (r-1)T)}{(1-T)\theta}, \quad \forall \ \theta \in [0, 1], \tag{19}$$

and then get

$$\underline{\theta} = \frac{k(r-1)T + (r - (r-1)T)}{(1-T)\alpha} \quad \text{and} \quad \frac{\partial\underline{\theta}}{\partial k} > 0. \tag{20}$$

We now illustrate this special case in Figure 2.

First, consider that the schedule e is originally set at a constant k^B. Substituting k^B into k in (19), one can obtain a downward-sloping loan rate schedule, such as R^B. Substituting k^B into k in (20), one can also obtain $\underline{\theta}^B$ such that a bank will accept any loan of quality $\theta \geq \underline{\theta}^B$ and reject all the others. When the regulator shifts the schedule e from k^B to another constant k^A, the resulting loan rate schedule becomes R^A, which is above R^B for all θ. The minimum acceptable quality then increases from $\underline{\theta}^B$ to $\underline{\theta}^A$, implying that a bank would now additionally reject loans of quality $\theta \in [\underline{\theta}^B, \underline{\theta}^A)$.

Recall that provision of risk-free deposits to the economy without incurring uncompensated liabilities for the regulator, and controlling banks' asset choices, are viewed as the two policy goals of bank regulations. Having successfully dealt with the first goal in Section 4, we are now in a position to discuss the second goal. If the regulator explicitly incorporates the second goal, its problem may now be restated as follows: select $\underline{\theta}$ and find $\{e(\theta), \pi(\hat{\theta})\}$ for all $\theta \in [\underline{\theta}, 1]$ that satisfies the conditions (12), (16) and

$$e(\underline{\theta}) = \frac{\alpha(1-T)\underline{\theta} - (r - (r-1)T)}{(r-1)T}. \tag{21}$$

Condition (21) gives a boundary condition for the schedule e and ensures that individual banks only grant loans of quality $\theta \geq \underline{\theta}$ and voluntarily reject all other loan applications.

The regulator's optimal choice of $\underline{\theta}$ may stem from the two-sided nature of its impacts on the loan market. That is, stricter capital standards would increase the competitive equilibrium loan rate schedule and, in turn, raise the minimum loan quality that banks find acceptable. This would tend to increase the "safety and soundness" of the banking system but would also shrink the aggregate credit supply in the economy.

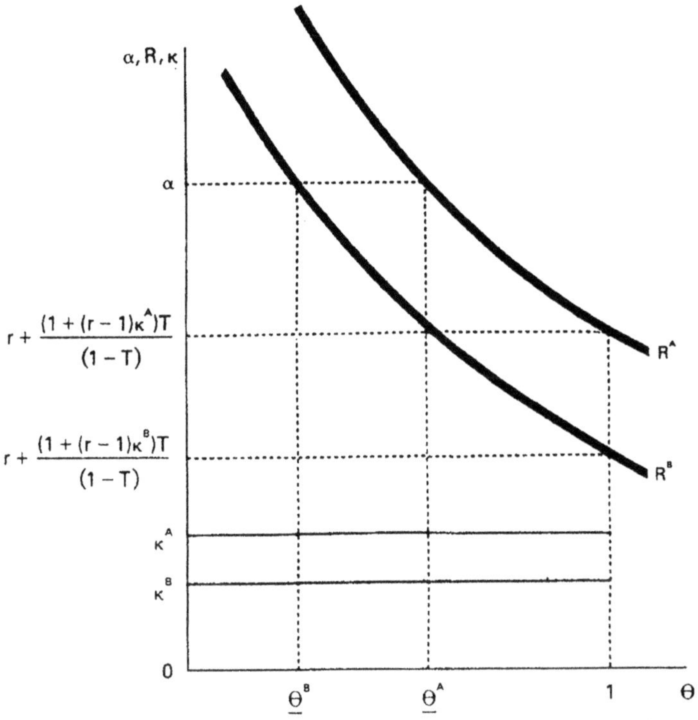

Note: Superscripts A and B denote that the associated variables correspond to before and after the shift in the uniform capital requirement schedule, respectively.

Figure 2. The impact of uniformly stricter capital requirements on a bank's lending behavior

This would have the undesirable macroeconomic effect of reducing aggregate investment and output. The degree of such recessionary impacts would, of course, depend on the extent to which firms rely on bank credit to finance their investments. The regulator must weigh these macroeconomic implications on social welfare to determine the socially optimal capital requirement schedule from an array of schedules that satisfy (16) and (21).

6. Conclusions

The recent problems in the U.S. banking industry have been largely attributed to the excessive risk-taking incentives banks have under a flat-rate deposit insurance system. Implementing risk-adjusted regulatory policies are especially difficult when regulators are less informed than banks about loan quality.

Our article shows that despite such asymmetric information problems, effective regulations can still be provided as long as the regulator can observe the interest rates banks charge in the competitive loan market. When deposit insurance is risk-adjusted, a capital requirement schedule that forces banks to charge higher loan rates for lower quality loans would guarantee incentive compatibility.

Such regulations would enable the regulator to induce banks to truthfully reveal their loan quality, levy risk-adjusted insurance premium, and control the minimum loan quality acceptable to banks in a competitive loan market. Such regulations would thus allow the deposit insurance system to break even, protect depositors from bank failures and also improve the safety of the banking industry.

Appendix

Proof of Lemma 1

Using (1) and (5), we rewrite $NPV^E(\theta, \hat{\theta})$ in (3) as

$$NPV^E(\theta, \hat{\theta}) = \frac{\theta[R(1 - T) - rD + (r - 1)DT]}{r} - (1 - \hat{\theta}D).$$

Partially differentiating this with respect to D, we get

$$\frac{\partial NPV^E(\theta, \hat{\theta})}{\partial D} = -\theta + \frac{\theta(r - 1)T}{r} + \hat{\theta} \gtreqless 0 \text{ iff } \hat{\theta} \gtreqless \theta^*.$$

The signs follow from the definition of θ^*. Here, the positive sign implies that E will be reduced to its required minimum, $e(\hat{\theta})$, and hence we get $D = (1 - e(\hat{\theta}))/\hat{\theta}$ using (1) and (5). The negative sign implies that $D = 0$ and $E = 1$. When the derivative equals zero, the bank becomes indifferent to its financing mix. That $\theta^* < \theta$ follows from $r > 1$. □

Proof of Proposition 1

We only show the first result. The second result then immediately follows from Lemma 1. Let us assume that the bank reports $\hat{\theta} \leq \theta^*$. Then the bank's current equity holders earn

$$NPV^E(R < R(\theta), E < 1) \leq NPV^E(R < R(\theta), E = 1)$$
$$< NPV^E(R = R(\theta), E = 1).$$

The first relationship is a restatement of Lemma 1 for $\hat{\theta} \leq \theta^*$ (the equality holds for $\hat{\theta} = \theta^*$). And the second relationship holds since NPV^E increases in R. Clearly, $NPV^E(R = R(\theta), E = 1)$ is the maximum value that the equityholders can earn (if $\hat{\theta} \leq \theta^*$), since R is bounded above by $R(\theta)$.

If the bank instead reports $\hat{\theta} > \theta^*$ and applies $R = R(\theta)$, then the equity holders earn

$$NPV^E(R = R(\theta), E = 1) < NPV^E(R = R(\theta), E = e(\hat{\theta})).$$

This inequality follows from Lemma 1 and establishes that the bank will report $\hat{\theta} > \theta^*$. □

Proof of Lemma 2

Differentiating $R(\theta)$ in (11) and rearranging the terms, we get

$$\frac{\partial R(\theta)}{\partial \theta} = \frac{\theta(r-1)Te' - [r-(1-e)(r-1)T]}{(1-T)\theta^2} \lesseqgtr 0 \quad \forall \quad \theta \in \Theta \quad \text{iff} \quad e' \lesseqgtr Q(\theta).$$

Since $T \in (0, 1)$ and it must be that $\theta > 0$, we obtain the signs from the definition of $Q(\theta)$. We get $Q(\theta) > 0$ because $e \leq 1$ and $r > 1$. □

Proof of Proposition 2

We first consider two loan qualities θ and $\hat{\theta}$, where $\theta < \hat{\theta}$. We consider two alternative cases.

Case 1: $R(\theta) > R(\hat{\theta})$. (a) Suppose that a bank making a loan of quality θ falsely reports $\hat{\theta}$. This bank is required to apply $R(\theta, \hat{\theta}) = R(\hat{\theta})$ and employ $D(\hat{\theta})$ and $E(\hat{\theta})$ given by (6). Thus, we get

$$\begin{aligned} NPV^E(\theta, \hat{\theta}) &= \frac{\theta[R(\hat{\theta})(1-T) - (r-(r-1)T)D(\hat{\theta})]}{r} - E(\hat{\theta}) \\ &< \frac{\hat{\theta}[R(\hat{\theta})(1-T) - (r-(r-1)T)D(\hat{\theta})]}{r} - E(\hat{\theta}) \\ &\equiv NPV^E(\hat{\theta}) = 0. \end{aligned}$$

The inequality follows from $\theta < \hat{\theta}$, since the sign of the bracketed term in the numerator must be positive. The last equality follows from the definition of $R(\hat{\theta})$ given by (11). Thus, such a bank would truthfully report θ. (b) Reversing the roles of θ and $\hat{\theta}$, we now suppose that a bank making a loan of quality $\hat{\theta}$ falsely reports θ. Since this bank is required to apply $R(\theta) > R(\hat{\theta})$ to the loan of quality $\hat{\theta}$, it will lose the loan applicant. Thus, such a bank would also truthfully report $\hat{\theta}$.

Case 2: $R(\theta) \leq R(\hat{\theta})$. (a) Consider the same bank as in Case 1a. If $R(\theta) < R(\hat{\theta})$, then this bank is required to apply $R(\hat{\theta})$, and hence it will lose the loan applicant. Thus, it would truthfully report θ. If $R(\theta) = R(\hat{\theta})$, then this bank is required to apply the same loan rate but will earn $NPV^E(\theta, \hat{\theta}) < NPV^E(\hat{\theta}) = 0$, where the inequality follows from $\theta < \hat{\theta}$. Thus, it would truthfully report θ. (b) Again reversing the roles of θ and $\hat{\theta}$, we consider the same bank as in Case 1b. This bank is required to apply $R(\theta) \leq R(\hat{\theta})$ to the applicant, and employ $D(\theta)$ and $E(\theta)$ given by (6). Thus, we get

$$NPV^E(\hat{\theta}, \theta) = \frac{\hat{\theta}[R(\theta)(1 - T) - (r - (r - 1)T)D(\theta)]}{r} - E(\theta)$$
$$< \frac{\theta[R(\theta)(1 - T) - (r - (r - 1)T)D(\theta)]}{r} - E(\theta)$$
$$\equiv NPV^E() = 0.$$

The inequality follows from $\hat{\theta} > \theta$. Thus, such a bank would falsely report $\theta < \hat{\theta}$.

We now generalize this result to state the proposition. Take any two values $\theta \in \Theta$ and $\hat{\theta} \in \Theta$ with $\theta < \hat{\theta}$. If $R(\theta) > R(\hat{\theta})$, then a bank would report truthfully. On the other hand, if $R(\theta) \leq R(\hat{\theta})$, then a bank with $\hat{\theta}$ would falsely report θ. Thus, it follows that if $R(\theta)$ is decreasing in θ for all $\theta \in \Theta$, then all banks would truthfully report. On the other hand, since a false report will be made if $R(\theta) \leq R(\hat{\theta})$ for some $\theta < \hat{\theta}$, truthful reports by all banks will be made if $R(\theta)$ is decreasing in θ for all $\theta \in \Theta$. Condition (16) then provides the result using Lemma 2. □

Proof of Proposition 4

Multiplying the RHS of equation (10) by r, letting $\theta = \underline{\theta}$, $R(\theta) = \alpha$, and $e(\theta) = m(\theta) + n$ for all $\theta \in [0, 1]$, and rearranging terms, we write

$$H(\underline{\theta}, n) = \underline{\theta}\, \alpha(1 - T) - (r - 1)T(m(\theta) + n) - (r - (r - 1)T).$$

By fixing $m(\theta)$ and applying the implicit function theorem to $H(\underline{\theta}, \text{n})$, we get

$$\frac{\partial\theta}{\partial n} = -\frac{\partial H/\partial n}{\partial H/\partial\underline{\theta}} = \frac{(r - 1)T}{\alpha(1 - T)} > 0.$$

The sign yields $\underline{\theta}^B < \underline{\theta}^A$. Proposition 3 then gives rise to the following result. A bank considering a loan of quality $\theta \in [\underline{\theta}^A, 1]$ would still make the loan, and hence its riskiness would remain unchanged. However, a bank considering a loan of quality $\theta \in [\underline{\theta}^B, \underline{\theta}^A]$ would forego the loan, potentially replacing it with another loan of quality $\theta \in [\underline{\theta}^A, 1]$. The bank's riskiness would decrease since it is inversely related to θ. □

Acknowledgments

We greatly appreciate comments from the editor, Henri Loubergé, and an anonymous referee. We also thank Angelo Baglioni, Arnoud Boot, Eric Briys, Yuk-Shee Chan, Andrew Chen, Mark Flannery, Ron Giammarino, E. Han Kim, Tom Lindley, D. Michael Long, Helena Mullins, S. Nagarajan, and seminar participants at the 1995 Seminar of the European Group of Risk and Insurance Economists, in Geneva, Switzerland, the 1992 European Finance Association meeting in Lisbon, Portugal, the 1992 Western Finance Association meeting in San Francisco, the 1991 Financial Management Association Meeting in Chicago, and the 1991 Northern Finance Association (NFA) conference in Montreal for their helpful comments. The authors thank the NFA and the Canadian Bankers Association for the Best Paper on Financial Institutions Award received at the 1991 NFA meeting. Mazumdar gratefully acknowledges an SSHRC grant and an FCAR grant that provided partial financial support for this project. The usual disclaimer applies.

Notes

1. See, for example, Kane [1986] for a proposal of six complementary reforms and Mester [1990] and Black [1991] for reviews. The most recent piece of legislation aimed at U.S. bank regulatory reform was the Federal Deposit Insurance Corporation Improvement Act (FDICIA) passed in 1991. In addition to risk-adjusted capital and deposit insurance premia the Act restricts loans from the Federal Reserve's discount window to meet banks, liquidity needs and introduces specific early closure rules for capital-deficient banks. See Mazumdar [forthcoming], and Chen and Mazumdar [1994, 1995] for a discussion of the moral hazard incentives associated with the discount window. See Chen and Mazumdar [1994] and references listed therein for a discussion of this Act.
2. For example, Besanko and Kanatas [1990] show that incentive compatible regulations could exist when a bank's management could improve a loan's cash flow by increasing managerial efforts that are unobservable by the regulator. Giammarino, Lewis, and Sappington [1990] also obtain a similar result in the presence of a bank's moral hazard behavior of directing resources towards nonproductive activities.
3. They then show that deposit insurance subsidy could provide incentive compatibility if capital requirement is higher and insurance premium is lower for a bank with better loan quality. It should be noted that Chan, Greebaum, and Thakor [1992] implicitly assume that the regulator does not observe bank loan rates.
4. The regulator's observation of these loan rates represents a first step toward bank monitoring, which partially constitutes the implicit regulations pointed out by Buser, Chen, and Kane [1981].
5. The competitive market assumption is critical to establish incentive compatibility, as we demonstrate later. Although such an assumption is standard in the literature, we should point out that if banks have monopoly power in the loan market, then interest rates may be determined by such monopoly power in addition to the firm's default risk. The lending rate may then not be a sufficient statistic to infer loan quality as suggested in this paper. Banks could acquire such monopoly power if information acquisition about borrowers was costly for lenders. Banks could then extract ex-post monopoly rents from their borrowers after having acquired such valuable information, thereby making it costly for firms to switch between lenders (see for example, Rajan [1992]). We do not model such additional complexities in this paper.
6. For an analysis of bank capital structure that also incorporates personal taxes, see Orgler and Taggart [1983].
7. By appealing to the revelation principle (e.g., Myerson [1979]), the regulator can restrict itself to incentive compatible regulations that would lead banks to truthfully reveal their privately observed loan qualities.
8. We thank the referee for highlighting these issues.
9. Mazumdar and Yoon's [forthcoming] model differs from ours insofar as they examine a scenario in which the bank monitors its client and can thereby improve the success probability of the borrower's project whereas we abstract from such bank monitoring issues in this article.
10. Whether or not a bank pays its deposit insurance premium using deposits depends on whether or not it is allowed. Disallowing such a possibility is equivalent to requiring additional capital.
11. The banks' equity market suffers from an Akerlof-type market failure as no outsiders can observe the banks' true loan qualities. Thus, the current equityholders, in equilibrium, prefer (or are indifferent) to contribute their own additional capital to the loan instead of raising outside equity capital. However, if all banks truthfully reveal their true loan qualities to the regulator and the market in equilibrium, then current shareholders would become indifferent to the two equity financing methods.
12. For a dollar deposit, the former is $\theta(r - 1)T/r$ and its sign is positive, while the latter is given by $-\theta + (1 - \pi(\hat{\theta})) = -\theta + \hat{\theta}$ and its sign depends on $\hat{\theta} \gtreqless \theta$. For derivation of these two terms see the proof of Lemma 1 in the appendix.
13. Kanatas [1986] provides an example of incentive compatible bank regulatory policies incorporating both deposit insurance and discount window.

14. Note that $R(\cdot, \cdot)$ corresponds to R in the previous section, where the arguments of $R(\cdot, \cdot)$ had been subsumed for simplicity.
15. To complete our argument let us consider the case when $R(\theta)$ is constant. In this case too, a bank with θ would falsely report $\hat{\theta} < \theta$ for the same reason as above. However, such a bank would again not falsely report $\hat{\theta} > \theta$ because of the higher capital requirement associated with reporting $\hat{\theta}$, which causes $R(\theta)$ to be constant.
16. As an extreme but interesting case, let us consider a laissez-faire economy in which $e(\theta) = 0$ for all θ, $T = 0$, and all deposits are uninsured. As all banks become indifferent between equity financing and deposit financing, competition in the loan market would yield a downward sloping $R(\theta)$. The loan rates banks charge would then become fully revealing.
17. See Pennacchi [1987] for empirical evidence related to this issue.
18. Thus our analysis may be interpreted as a restatement of classical finance theory concerning the risk-return tradeoff for bank loan contracts under asymmetric information.
19. Note that in the usual argument for the moral hazard behavior it is assumed that the bank set value remains unchanged when its variance parameter increases. In our model, however, as the variance of the loan project goes up (inversely with the quality), its value goes down.
20. The term *riskiness* is used to represent the asset side risk which is ultimately shared by bank equityholders and the deposit insurer.
21. Early papers by Kahane [1977], Koehn and Santomero [1980], and Kim and Santomero [1988] argue that stricter risk-independent capital standards, without accompanying portfolio constraints, would not reduce a bank's risk-taking incentive. However, Keeley and Frulong [1990] lend support to the conventional wisdom.

References

BESANKO, D., and KANATAS, G. [1990]: "Bank Capital Regulations: Time Consistency, Hedging and Incentive-Compatibility," Working Paper, Indiana University.

BLACK, R.P. [1991]: "Reflections on Deposit Insurance," *Economic Review*, FRB of Richmond, (January/February), 3–6.

BUSER, S.A., CHEN, A.H., and KANE, E.J. [1981]: "Federal Deposit Insurance, Regulatory Policy, and Optimal Bank Capital," *Journal of Finance*, 35, 51–60.

CHAN, Y.-S., GREENBAUM, S.I., and THAKOR, A. [1992]: "Is Fairly Priced Deposit Insurance Possible?," *Journal of Finance*, 47, 227–245.

CHEN, A.H., and MAZUMDAR, S.C. [1994]: "The Impact of Regulatory Interactions on Bank Capital Structure," *Journal of Financial Services Research,* 8, 283–300.

CHEN, A.H. and MAZUMDAR, S.C. [1995]: "The Regulatory Safety Net and Bank Moral Hazard Behavior: A Contingent-Claims Analysis," *Research in Finance*, 12, 115–140.

FLANNERY, M.J. [1989]: "Capital Regulation and Insured Banks' Choice of Individual Loan Default Risks," *Journal of Monetary Economics*, 24, 235–258.

GIAMMARINO, R.M., LEWIS, T.R., and SAPPINGTON, D.E.M. [1990]: "An Incentive Approach to Banking Regulation," *Journal of Finance*, 48, 1523–1542.

JAFFEE, D.M. [1989]: "Symposium on Federal Deposit Insurance for S&L Institutions," *Journal of Economic Perspectives*, 3, 3–9.

KAHANE, Y. [1977]: "Capital Adequacy and the Regulation of Financial Intermediaries," *Journal of Banking and Finance*, 1, 207–217.

KANATAS, G. [1986]: "Deposit Insurance and the Discount Window: Pricing Under Asymmetric Information," *Journal of Finance*, 41, 437–450.

KANE, E. [1986]: "Appearance and Reality in Deposit Insurance: The Case for Reform," *Journal of Banking and Finance*, 10, 175–188.

KEELEY, M.C., and FURLONG, F.T. [1990]: "A Reexamination of Mean-Variance Analysis of Bank Capital Regulation," *Journal of Banking and Finance*, 14, 69–84.

KIM, D. and SANTOMERO, A.M. [1988]: "Risk in Banking and Capital Regulation," *Journal of Finance*, 43, 1219–1233.

KOEHN, M. and SANTOMERO, A.M. [1980]: "Regulation of Bank Capital and Portfolio Risk," *Journal of Finance*, 35, 1235-1244.
MAZUMDAR, S.C. [forthcoming]: "Bank Regulations, Capital Structure and Risk," *Journal of Financial Research*.
MAZUMDAR, S.C., and YOON, S.H. [forthcoming]: "Loan Monitoring, Competition and Socially Optimal Bank Capital Regulations," *Journal of Risk and Insurance*.
MERTON, R.C. [1977]: "An Analytic Derivation of the Cost of Deposit Insurance and Loan Guarantees," *Journal of Banking and Finance*, 1, 3-11.
MESTER, L.J. [1990]: "Curing Our Ailing Deposit-Insurance System," *Business Review*, FRB of Philadelphia (September/October), 13-24.
MYERSON, R. [1979]: "Incentive Compatibility and the Bargaining Problem," *Econometrica*, 47, 61-74.
ORGLER, Y.E. and TAGGART, R.A. JR. [1983]: "Implications of Corporate Capital Structure Theory for Banking Institutions," *Journal of Money, Credit and Banking*, 15, 212-221.
PENNACCHI, G.G. [1987]: "A Reexamination of the Over- (or Under-) Pricing of Deposit Insurance," *Journal of Money, Credit, and Banking*, 19, 341-360.
RAJAN, R. [1992]: "Insiders and Outsiders: The Choice Between Informed and Arm's Length Debt," *Journal of Finance*, 47, 1267-1400.

GPSR Compliance

The European Union's (EU) General Product Safety Regulation (GPSR) is a set of rules that requires consumer products to be safe and our obligations to ensure this.

If you have any concerns about our products, you can contact us on ProductSafety@springernature.com

In case Publisher is established outside the EU, the EU authorized representative is:

Springer Nature Customer Service Center GmbH
Europaplatz 3
69115 Heidelberg, Germany

Batch number: 10823959

Printed by Printforce, the Netherlands